Penn State University's

Introduction to Research in

Astronomy &

Planetary Sciences

2nd Edition

Penn State University's

Introduction to Research in
Astronomy &
Planetary Sciences

Stephanie J. Slater
*CAPER Center for Astronomy &
Physics Education Research*

Timothy F. Slater
University of Wyoming

Christopher Palma
Penn State University

Julia Kregenow
Penn State University

2nd Edition

Cover Photo: *Penn State's new CDK 24" telescope from PlaneWave Instruments, located in one of the 3 domes on the roof of Davey Lab on the Penn State University Campus*

First Printing of 2nd Edition: January, 2016

Second Printing of 2nd Edition: August, 2016

ISBN-13: 978-1515190530

ISBN-10: 1515190536

CAPER Center for Astronomy &
Physics Education Research
604 So. 26th St., Laramie, WY 82070 USA
www.caperteam.com

After decades of hanging out in science classrooms and labs, we have learned one thing for certain: science classes do a pretty good job of sucking all of the joy out of learning science. This is true even in courses like astronomy, which by our reckoning, could be, and should be, the most interesting science course anyone ever takes. Our take on this is that, without full intent, science instructors and textbooks do a lousy job of getting across what science is about, what kinds of talents are needed in science, and how exactly the science is done.

We can't change all of that in one skinny little lab manual, so it is our intention to go after one piece of the business: the issue of how science is really done, astronomy in particular.

To begin we would ask you to imagine every science classroom that you've ever been in, and every science textbook that you've ever been asked to read. Most likely, your instructor, with all good intentions, put on display, a poster of "The Scientific Method." Likewise, nearly every science textbook begins with a dreadful chapter on "The Scientific Method." We hate these things. They're horrible. We do this stuff for a living so you would think that we're immune to the anxiety that such things inspire, but we are not. "The Scientific Method," to us anyway, feels rigid and joyless and full-on lacking in creativity. But worse than that, in our many years in science, we have learned that the seven-or-so "Scientific Method" that we were all required to memorize in seventh grade, is nothing less than a big 'ol, fat lie.

Don't get us wrong. Science is absolutely the kind of thing where we try to go about making our claims in systematic ways, but that doesn't mean that it's done using the Scientific Method (SM). Many, perhaps most, fields in science, do not rely on work that is experimental, or that involves hypotheses. In astronomy for example, just how is one to experiment on a galaxy? And we've heard astronomers who have been in the business for 30 years say that they haven't written a hypothesis since they were in high school biology! The truth is that there are many, many methods that one can use to do science, and nearly all of the methods that are really used by scientists are more flexible, and more interesting than we were ever taught in school.

So there will be no Scientific-Methoding in this lab manual. Instead we will focus on the ways that real scientists actually approach their work, the practices that they engage in, and in most cases, the actual data that they use to do modern astronomy. In this lab manual, we are going to focus on the act of doing science in astronomy, which is really just about three things:

Questions
Data
Evidence based claims

General Information about these activities:

This book contains a collection of astronomy assignments like no other book available. The lessons reflect an innovative approach to learning astronomy by putting you, the learner, at the center of each and every lesson. In these lessons, you decide what specific topics you want to study, create your own research questions, design your own strategies to pursue evidence, and defend your scientific conclusions based on the data that you collect. If this sounds like you are responsible for your own learning in these lessons, you are exactly right. In *this book* you are the astronomer out there collecting data about objects in the cosmos.

These lessons use a carefully structured in a multi-step approach to help you learn how astronomers conduct their trade. But don't worry; we won't just turn you loose without any direction. In these lessons, during your first experience with inquiry, you are guided through the entire scientific inquiry process, from given research questions to the appropriate content and format for a scientific conclusion. Then, in your second experience, you will generate your conclusions independently, with the previous experience set out as a guide for content and format. This will help you make sense of astronomical data that has been purposefully planned, collected, and analyzed with the guidance of your instructor. You will first construct and defend conclusions based upon data that is, provided for you. By the time you reach your third inquiry, you have been exposed to two experiences in which you were guided through the process of data collection and analysis. During this third inquiry data collection and analysis becomes an independent task. By the fourth inquiry, you will have received explicit instruction on the connection between the research questions or hypotheses, and the procedure undertaken to address them three times. By then, you will be prepared to take responsibility for creating a plausible method for collecting data given a research prompt. By the fifth inquiry, you will have now seen four examples of quality research questions/hypotheses, and their relationship to procedures, data collection and conclusions. At this point you will be positioned to successfully conduct an entire inquiry cycle in astronomy. This strategy is specifically designed to provide you with repeated success in doing science and a sense of how the pieces of the scientific process connect to each other.

We know that astronomy might initially sound like a far-out science, and indeed it is in many ways. To help you learn how astronomy is done, we welcome you to engage in these lessons and begin to see the Universe as an astronomer does – as a wonderful and fascinating world in which to pursue questions of your own choosing. We invite you to engage in astronomical inquiry.

Table of Contents

DE	Designing an Experiment

Background, Relevance, and Motivation for the Assignment: There are many times in life when someone makes a statement about the effectiveness of a product or about a certain condition in society. How do you know if such statements are correct? In order to decide if a claim is reasonable or not, one needs to examine all the relevant evidence used to make the claim. This is exactly what we do in science: making and investigating claims about why a certain condition exists in nature. One valuable tool for gathering evidence is to design an experiment. In this activity we are going to explore ways to design an experiment that will provide convincing evidence to support or reject a given claim.

This assignment is based on an activity written by Bruce Wellman for the POGIL project, and is used here with permission by the author.

Goal: The purposes of this assignment are to
- empower you to critically evaluate claims that are made in everyday life, and realize you can apply scientific reasoning to answer them,
- introduce you to the types of questions and style of activity we will be using throughout this course so you get used to thinking about these issues and practicing these skills, and
- do so using nonthreatening, commonplace real-life examples with which you are comfortable, so you get a sense for what these activities are asking for before we throw them into the context of (probably) unfamiliar astronomy topics.

Part I: The iPod Question

Part II: Two possible experiments to test the question

	Friend A's Experiment	**Friend B**'s Experiment
Beginning Question	Does the color of an iPod affect the length of playtime on one charge?	Does the color of an iPod affect the length of playtime on one charge?
Procedure & Tests	One red iPod and one silver iPod were charged fully and then used until the battery was completely dead.	Three red iPods and five silver iPods were borrowed from various friends and charged completely. Each iPod was loaded with the song "Thriller" (by Michael Jackson). The song was played over and over again (on repeat) and the length of time the iPod was able to play the music was recorded.
Data & Observations	See table below	See table below
Claims	The color of an iPod impacts how long the MP3 player will run on one charge and the red iPods last longer than other colored iPods.	The color of the iPod does not impact how long the MP3 player will run on one charge.

Friend A's Data & Observations:

	Red iPod	**Silver** iPod
play back time	5 days	2 days

Friend B's Data & Observations:

	Trials			
iPods	**Run 1**	**Run 2**	**Run 3**	*Average of 3 runs*
Red A	855 min	823 min	875 min	*851 min*
Red B	858 min	873 min	819 min	*850 min*
Red C	890min	799 min	808 min	*832 min*
Silver A	788 min	815 min	879 min	*827 min*
Silver B	906 min	870 min	815 min	*864 min*
Silver C	808 min	902 min	850 min	*853 min*
~~Silver D*~~	~~640 min~~	~~760 min~~	~~650 min~~	*~~683 min~~*
Silver E	866 min	832 min	857 min	*852 min*

*Note: The times for iPod Silver D were not used in this comparison because they seem to be outside the range of the other identical iPods. I assumed there must be something wrong with that specific iPod.

> Average of all RED iPods = 844 min
> Average of all SILVER iPods = 849 min

Write down some descriptive words (NO MORE than 8 *total*) which summarize your group's first impressions of each of the two experiments.

1) Friend A's Experiment	2) Friend B's Experiment

3)   a. How many iPods were tested by Friend A?

           b. How many iPods were tested by Friend B?

4)   a. How many **times** did Friend A test each iPod?

           b. How many **times** did Friend B test each iPod?

5)   By what process did Friend A "measure" the battery life of the iPod?

6)   By what process did Friend B "measure" the battery life of the iPod?

7)   Which of the two methods for "measuring" the battery life of the iPod provides the most reliable measure of the battery life? Explain your reasoning.

8)   Brainstorm a list of at least five possible reasons why the red iPod in Friend A's experiment played music longer than the silver iPod. Be creative, yet practical. Consider as many reasons as possible.

9)   Which, if any, of the problems your group listed in the previous question were minimized by Friend B's choice to test each iPod by playing the same song over and over? Provide reasons for your answers (or some reasons for using the same song over and over).

10) Which, if any, of the problems your group listed in question #8 were minimized by Friend B's choice to test multiple iPods of each color? Provide reasons for your answers (or some reasons for testing multiple iPods of each color).

11) In order to have a valid and convincing claim derived from evidence gathered from an experiment, why must a person identify ALL possible conditions that could change the outcome of the experiment and then only vary ONE condition?

12) Looking at Friend B's experiment, if she had only done her **first** trial of data (**Run 1**) (including Silver iPod D), would she have made the same conclusion that Friend A made? Support your answer with numerical evidence.

13) Looking at Friend B's experiment, if she had only done her **second** trial of data (**Run 2**) (including Silver iPod D), would she have made the same conclusion that Friend A made? Support your answer with numerical evidence.

14) What information became apparent after Friend B performed **3 trials** of the battery length on many different iPods with the same color that was not visible by only doing **1 trial** on different iPods with the same color?

Notice that Friend B included the data from the Silver D iPod on her final data table even though she chose not to use that data in calculating the averages. She felt justified in excluding the Silver D data because it was consistently and substantially lower than all the other measurements, but it was crucial that she include the data in the table: it allows others to follow her thinking and to re-calculate the averages using the excluded data to see for themselves if her decision not to include those data was reasonable. Excluding data as outliers is often justified. But it is never justified to *omit* or *hide* data. A scientist who does this will quickly find him or herself out of a job, and justifiably so, because hiding (or falsifying) data is unethical, unprofessional, dishonest, and counter to the purpose of science.

15) When designing an experiment, why would performing multiple trials at each condition be a wise choice? Give TWO reasons.

16) Friend B concluded that the playing times of the two different colored iPods were essentially the same (only varied by 5 min on average). Do you agree or disagree with her conclusion? Explain your reasoning.

17) If a different experiment were performed and the researcher found that two different conditions produced playing times which also varied by 5 minutes, but the average playing times for each iPod were 17 min and 22 minutes respectively, would you make the same conclusion? Why or why not?

18) When making a decision to accept or reject a claim based upon experimental evidence, does the numerical difference in outcomes (i.e. SUBTRACTING two results to compare them) provide enough information? Why or why not?

Part III: Another example experiment

19) A student proposed the following question for the class to investigate: **"Does the temperature of a tennis ball affect how high the ball will bounce?"** Examine the possible table of data at right.

	Trial 1	Trial 2	Trial 3
10° C	15.3 cm	14.2 cm	16.1 cm
15° C	14.5 cm	13.8 cm	17.9 cm
20° C	15.8 cm	7.2 cm	14.8 cm
25° C	14.6 cm	16.2 cm	15.7 cm

Write a claim which can be made from these data. Be sure to explain your reasoning, and support your answer quantitatively.

Often the most challenging part of scientific experiments is designing them in the first place so that they are likely to provide reliable results. We will be practicing this skill throughout this entire course, taking a "backwards" approach each week: you will begin by practicing the later stages first (analyzing and interpreting data, drawing conclusions), then the intermediate stages (collecting the data, designing the procedure to get the data), and ultimately starting from the very beginning (formulating the research question and setting the scope of the experiment).

Previously in this assignment, you have already gotten some practice analyzing and interpreting data that was gathered to investigate a question, and considering what variables are involved in setting up an experiment to provide those data, and critically analyzing an experimental setup. Now, your final tasks will be to start at the very beginning of the process of experimental design:
(A) **Designing a step-by-step experimental procedure** that is likely to yield useful results, and is sufficiently detailed to show other researchers exactly what you did to get those results, such that they could repeat it without you if necessary,
AND, even before that,
(B) **Posing an answerable research question** that you could actually pursue an answer to. This is almost an art, formulating a question that is answerable with data you can readily gather, but is not too simple as to be trivial, and also not too difficult as to be prohibitively onerous with the tools available.

In subsequent activities throughout the rest of this Astro 11 course, you will actually be gathering the data for yourself to answer research questions about astronomy. But for the remaining few questions in THIS activity, you will not actually gather the data. Proceed in your experimental design pretending that somebody else will gather them, and assuming they have the necessary skills and resources.

Part IV: Design your own procedure

20)   Imagine you are tasked to investigate the impact of drinking diet sodas sweetened with aspartame and short-term memory function. Describe precisely what evidence you would need to collect in order to answer the research question of, **"How does the amount of aspartame consumed per day correlate with the ability to remember names of strangers shortly after meeting them?"** Obviously, you will not actually complete the steps in the procedure you are writing.

Create a detailed, step-by-step description of evidence that needs to be collected and a complete explanation of how this could be done - not just "record how much aspartame was consumed and count how many names were remembered", but exactly what would someone need to do to accomplish this. You might include an empty data table the researchers could fill in - the goal is to be precise and detailed enough that someone else could follow your procedure without you on hand to ask for clarification. Do NOT include generic nonspecific steps such as "analyze data" or "present conclusions" -- these are meaningless filler. Be specific!

Tip: *Be specific enough to define your methods, but you don't have to go overboard on implementation details. You can assume that any researcher carrying out these steps will be intelligent and can make reasonable decisions and judgment calls. For example, your procedure might include something like, "Use one specific kind of aspartame-sweetened soda: diet caffeine-free Pepsi in 12-oz servings. Give 5 participants each 2 sodas per day, 5 each 4 sodas per day, 5 each 6 sodas per day, and 5 each no sodas per day". You do NOT need to specify bottles vs. cans of soda, how fast they drink it, or what they do the rest of the day. However, you SHOULD specify things like "Make sure the participants are not consuming any other sources of aspartame in their diet" in order to keep that variable controlled.*

Part V: Devise your own research question

21)   Your next task is to design an answerable research question on a new topic: something related to **amount of television watched and its effects on the body or mind**. (This is deliberately vague to allow you creativity to think of your own question, and to force you to think about what makes a question answerable.) Some example questions to get you thinking about the possibilities (you CANNOT use these questions; you must think of your own): "Does number of hours of TV watched per week by college students negatively correlate with number of pages of books read per week?" or "Do people who watch more TV have a higher incidence of diabetes?" or "Do children who watch educational programming (e.g. Sesame Street) do better or worse in school than children who watch no TV at all?"

You do not have to write the procedure for how you would go about collecting the data to test this question.

Part VI: Summary and Reflection

22)   Create a 50-word summary, in your own words, that describes what makes a good experiment. You might choose to comment on some of the following: the research question, experimental setup, data collection, analysis, and interpretation. Base your answer on what you have done in today's activity, and not what you have learned in another class or elsewhere. Every word should convey something important and meaningful. There is no room for "filler" that doesn't serve a purpose.

Tip: 50 words is NOT MUCH. (The previous paragraph has 76 words. This one has 109.) **You will have to keep your answer very succinct and pithy.** Think of this as writing a concentrated answer, where every nonessential word has been removed. We will practice this every week. As you get better at it, strive to bring this approach to your other writing in other classes and in your life. Being concise is an invaluable skill in your education, future career, and even in communicating with people in your personal life. If you find yourself writing much more than 50 words, you have missed the point of this question.

Your Name: _____

Group Member(s): _____

Big Idea: The Hubble Space Telescope image known as the "Hubble Ultra Deep Field" reveals a variety of previously unknown objects in the very distant universe that can be systematically and scientifically counted, organized, and classified.

Computer Setup and/or Materials Needed:

a) Access the Hubble Ultra Deep Field image at the following URL:
    http://www.spacetelescope.org/static/archives/images/screen/heic0406a.jpg

b) Access the *SkyWalker* website at: http://www.aip.de/groups/galaxies/sw/udf/swudfV1.0.html

c) *Note: There is no expectation that students have studied galaxies prior to completing this research project.*

Phase I: Exploration

1) Access the online Hubble Space Telescope Image at the first link above. *You might be able to make it larger and smaller by "left clicking" on the image with your mouse.* Most of these objects are galaxies far, far from Earth. However, a few objects are nearby stars, as indicated by "four points" on the image, like shown at left.

How many stars can you find? _____

2) Again, most of the objects in the Hubble Ultra Deep Field image are not individual stars, but actually distant **galaxies**—*isolated collections of millions or billions of stars that look like a tiny dot or cloud.* Determine how many galaxies are found in the image. *Since counting each galaxy is not practical, one strategy for estimating the total number is to precisely count the number of galaxies in one small section of the image, then multiply the result by the appropriate number so that we have an estimate for the number of galaxies in the whole image. For example, if we counted the number of galaxies in 1/4 of the image, then we would multiply the result by 4 to find the approximate number of galaxies in the entire image (note that even this is impractical, as 1/4 of the image still contains too many galaxies to count one by one).* Keep in mind, every point of light that is not a star that you identified above is indeed a galaxy... even the smallest dots!

What is the total number of galaxies in this image? _____

Write one or two sentences describing how you arrived at this number and show all calculations:

3) Some of the galaxies are orange-red in color, while others are white and still others are blue. What is the most common color of galaxy in the image? *Precisely explain how you determined this, not just "I looked and saw more of this color."*

4) If we assume that all of the galaxies in this image have the same diameter, then the ones that are close appear larger and the ones that are more distant appear smaller. Are most of the galaxies in this image relatively near or relatively far? What is your evidence?

Phase II: Does the Evidence Match a Given Conclusion?

5) Access the interactive Ultra Hubble Deep Field site through the *SkyWalker* website at: http://www.aip.de/groups/galaxies/sw/udf/swudfV1.0.html

The green circle in the top left hand corner is a sort of "magnifying glass" that you can drag around that will let you look at close up portions of the Hubble Ultra Deep Field. *Note that the picture is about 8 green circles wide and 10 green circles tall, for a total of about 80 green circles over the whole image.*

Make rough sketches of the five <u>closest</u> galaxies you can find in the image, concentrating on the galaxy itself and not any background galaxies in the circle.

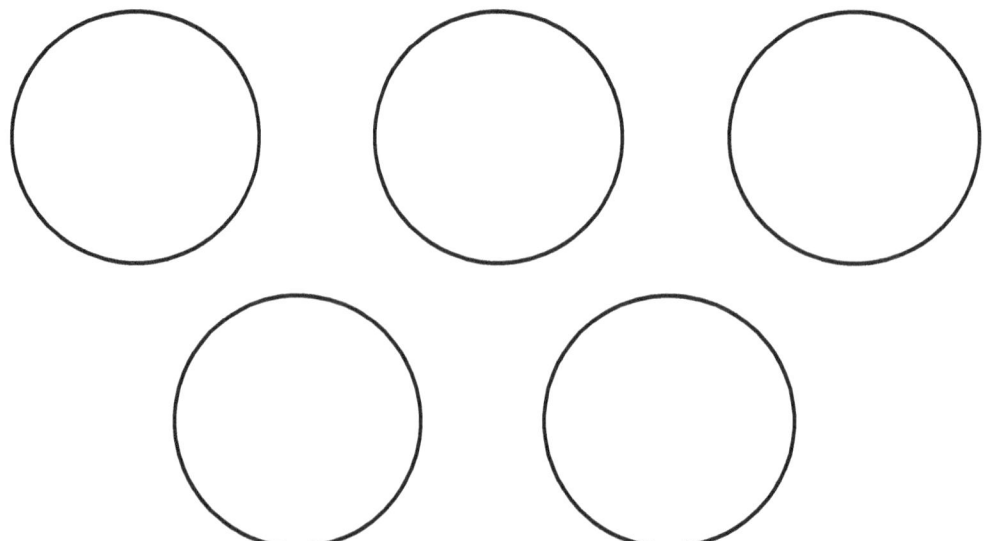

6) Consider the research question,"What is the most common type of nearby galaxy?" If a fellow student proposed a generalization that **"nearby galaxies are equally split between round, featureless galaxies and elongated, spiral shaped galaxies,"** would you agree or disagree with this generalization based on the evidence you collected above on the shapes of nearby galaxies? Explain your reasoning and provide specific evidence either from the above tasks or from new evidence you yourself generate using the SkyWalker website.

Phase III: What Conclusions Can You Draw from This Evidence?

The Hubble Ultra Deep Field is one of the most important images in astronomy because it shows some of the most distant galaxies in the universe. What conclusions and generalizations can you make from the following data collected by a student by randomly positioning the green circle in an effort to answer the question, **"What is the general distribution of galaxy colors?"** *Explain your reasoning and provide the specific evidence you are using, with sketches or pie charts or graphs if necessary, to support your reasoning.*

Green Circle Sample Number	Number of Red-Orange Galaxies	Number of Blue-White Galaxies
1	7	27
2	10	16
3	15	19
4	10	29
5	12	27

Data collected at http://www.aip.de/groups/galaxies/sw/udf/swudfV1.0.html

7) Evidence-based conclusion:

Phase IV: What Evidence Do You Need To Pursue?

Imagine your team has been assigned the task of writing a news brief for your favorite news blog about the differences between the numbers of nearby and extremely distant galaxies in the universe. Describe precisely what evidence you would need to collect, and how you would do it, in order to answer the research question of, **"Are there more nearby galaxies or more extremely distant galaxies?"** You do not need to actually complete the steps in the procedure you are writing.

8) Create a detailed, step-by-step description of evidence that needs to be collected and a complete explanation of how this could be done—not just "move the green circle around and look at how many big and how many small," but exactly what would someone need to do, step-by-step, to accomplish this. You might include a table and sketches—the goal is to be precise and detailed enough that someone else could follow your procedure.

Phase V: Formulate a Question, Pursue Evidence, and Justify Your Conclusion

Your task is to design an answerable research question, propose a plan to pursue evidence, collect data using the interactive *Ultra Hubble Deep Field* site (or another suitable source pre-approved by your lab instructor), and create an evidence-based conclusion about the characteristics of galaxies in our universe that we have not previously addressed.

Specific Research Question:

Step-by-Step Procedure to Collect Evidence:

Data Table and/or Results:

Evidence-based Conclusion Statement:

Phase VI: Summary

Create a 50-word summary, in your own words, that describes the characteristics and distribution of galaxies in our universe. You should cite specific evidence that you have collected in your description, not describe what you have learned in class or elsewhere. Feel free to create and label sketches to illustrate your response.

2	Observing the Sun's Position and Motion

Big Idea: Sky objects have properties, locations, and predictable patterns of movements that can be observed and described. Those motions explain such phenomena as the day, the year, the seasons, phases of the Moon, and eclipses.

Goal: Students will conduct a series of inquiries about the motion of the Sun in the sky using sky simulation software and learn how the Sun follows different pathways at different times of the year.

Computer Setup:

Open the program *Stellarium* on your lab computer. If you are working outside of the lab, you can download and install *Stellarium* for free from http://www.stellarium.org/.

Hover your mouse over the bottom-left hand side of the screen until a menu appears. Select the top button, "Location Window." This will allow you to set your location to anywhere around the world, or any of the planets (or Sun) in our solar system. For this lab, we will set our location to "State College, PA." In the search box (the one with the magnifying glass next to it) in the location window, type "State College" and select the location from the list (it should be the only one). Click "Use as default" (if not already selected) so that the location is set to State College for any future uses. Simply close the location window to save these settings. *Be sure when you've set State College as your location that the arrow on the map is pointing to State College. If it is not, then you have selected the wrong location.*

The time is set to the current time by default, so the sky you see on your screen should appear as it does right now outside (except for any weather). If you open the "Date/time" window, the time is shown in military time (i.e. 24 hour clock, so 6:00PM is represented as 18:00 and so on). We want to be able to see the constellations at any time of day, so hover over the left-hand menu and select "Sky and viewing options window." Uncheck the box that says "Show Atmosphere" and set the "Light Pollution" to "1." Both removing the atmosphere and reducing light pollution will allow us to see the sky without any residual light getting in the way. While these are not realistic viewing conditions, they will let us gather more information for this activity.

Note: Stellarium will show you the sky at your selected location, but will not change the time to the local time where you are observing. The time shown in the date and time window will always be in your computer's time zone. To correct this discrepancy, you should convert all of your given (system) times to the local time at your observing location.

Phase I: Exploration

1) On a map of the United States, north is towards the top of the page and west is to the left. Zoom out (either using your mouse scroll wheel or using "Ctrl+down" or "Cmd+down") until you can see the whole sky as a circle. In *Stellarium* and other sky charts, when north is towards the top of the page, west is to the *right*. Why are the directions of east and west flipped in a map of the sky compared to a map of the ground?

2) Return the sky to the default zoom settings. This is the current sky. Locate the Sun, and left click on it. You should see some basic information about the Sun, including its current position (Az/Alt) appear in the top-left corner of your screen. Select the checkboxes that show constellation boundaries and labels in "Sky and viewing options" under the "Markings" tab.

To which constellation is the Sun closest? *If you do not see the Sun,* Stellarium *may be set to nighttime. Add or subtract enough hours until the Sun appears.*

3) Increase the time one hour by selecting the "Date and Time window" along the left-hand menu and clicking the up arrow over the current hour. Both qualitatively and quantitatively, how has the Sun's position changed on the sky?

4) Slowly increase the time to later and later in the day. Determine exactly what time (hours and minutes) the Sun will set tonight.

Sunset: _____

5) To which constellation was the Sun closest when it set tonight? *Note: if the constellation name is hidden by the ground, you can make the ground invisible by un-selecting "Show Ground" under "View" in "Sky and Viewing Options." Alternately, you can change the scenery under "Landscape" in "Sky and Viewing Options" to remove the trees and buildings. When possible, it's helpful to keep the ground visible.*

6) Is this the same or different than where the Sun was earlier in the day?

7) What generalization can you make about the <u>relative speed</u> at which the Sun and the stars move throughout the sky over the course of a day?

8) What generalization can you make about the path and direction in which the Sun and stars move through the sky over the course of a day?

Set *Stellarium* to show the sky as it would be tonight at sunset. Pause the rotation of the sky by selecting the "Set normal time rate" button on the bottom menu. Zoom out until you can again see the whole sky as a circle, and orient the sky so that North is at the top of the screen. Make sure that you can still see the constellation lines and labels.

9) On what part of the map (left, right, top, bottom, or center) is the constellation that appears *highest* in the night sky (i.e. straight overhead)? What is the name of this constellation?

   Circle one: left | right | top | bottom | center   Name:

10) On what part of the map (left, right, top, bottom, or center) is the constellation that appears near the *northern* horizon? What is the name of this constellation?

   Circle one: left | right | top | bottom | center   Name:

11) On what part of the map (left, right, top, bottom, or center) is the constellation that appears near the *eastern* horizon? What is the name of this constellation?

   Circle one: left | right | top | bottom | center   Name:

Set *Stellarium* for THREE HOURS after sunset tonight.

12) On what part of the map (left, right, top, bottom, or center) is the constellation that now appears highest in the night sky (i.e. straight overhead)? What is the name of this constellation?

   Circle one: left | right | top | bottom | center   Name:

13) Where did the stars that used to be at this position move?

14) On what part of the map (left, right, top, bottom, or center) is the constellation that appears near the northern horizon? What is the name of this constellation?

   Circle one: left | right | top | bottom | center   Name:

15) Where did the stars that used to be at this position move?

16) On what part of the map (left, right, top, bottom, or center) is the constellation that appears near the western horizon? What is the name of this constellation?

Circle one: left | right | top | bottom | center   Name:

17) Where did the stars that used to be at this position move?

18) On what part of the map (left, right, top, bottom, or center) is the constellation that appears near the eastern horizon, where the Sun rises? What is the name of this constellation?

Circle one: left | right | top | bottom | center   Name:

19) Where did the stars that used to be at this position move?

20) *If* you were to change the time to midnight, predict what would be different about the positions of the stars.

21) What generalizations can you make about how the stars change position over the course of the night?

Phase II: Does the Evidence Match the Conclusion?

Consider the research question, **"How does the time of sunset change over the course of a year at this location?"** Using *Stellarium*, you will collect evidence to answer this question and use the information you gathered to challenge the conclusion proposed below.

In the data table below, fill in your observing location, the day of your observations for each month, and the corresponding sunset time each day.

Location: _____

Date (Month, Day, Year)	Time
January	
February	
March	
April	
May	
June	
July	
August	
September	
October	
November	
December	

22) A fellow student proposes the following generalization: "**Sunset time changes about one hour per month, setting earlier and earlier in the fall and then setting later and later in the spring.**" Based on the evidence you've collected in this section, do you agree or disagree with that statement? Explain your reasoning, citing evidence from this section or from additional evidence gathered using *Stellarium*.

Phase III: What Conclusions Can You Draw from the Evidence?

Most of us would agree that the Sun sets in the general direction of West. What conclusions and generalizations can you make from the following data collected by a fellow student to answer the question, **"How does the direction that the Sun sets change during the fall and winter months as observed from State College?"** *Explain your reasoning and provide evidence to support your reasoning.*

Date	Sunset Time (EST)	Azimuth
August 15, 2013	8:05 PM	288 (WNW)
September 15, 2013	7:17 PM	274 (W)
October 15, 2013	6:27 PM	258 (WSW)
November 15, 2013	4:48 PM	245 (WSW)
December 15, 2013	4:39 PM	238 (SW)

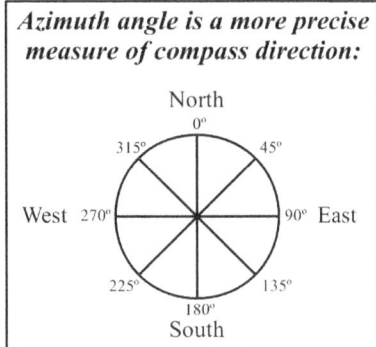

Azimuth angle is a more precise measure of compass direction:

23) Evidence-based conclusion:

Phase IV: What Evidence Do You Need?

Describe precisely what evidence you would need to collect in order to answer the research question of, **"How does the noon-time Sun's position above the southern horizon change over the course of a year?"** You do not actually need to complete the steps in the procedure you are writing.

24) Create a detailed, step-by-step description of the evidence needed and a complete explanation of how this could be done using *Stellarium*. Your procedure should be more than just "measure the position of the Sun," but exactly what someone completely unfamiliar with the program would need to do, step-by-step to accomplish this. This includes any non-default settings that are needed to collect the evidence, but *shouldn't* include meaningless filler like "analyze data and draw conclusions."

Phase V: Formulate a Question, Pursue Evidence, and Justify Your Conclusion

Your task is to design an answerable, original research question about some motion or position of the Sun in the sky that we have not previously addressed, propose a plan to pursue evidence, collect data using *Stellarium* by following your outlined procedure, and create an evidence-based conclusion that answers your research question.

Specific Research Question:

Step-by-Step Procedure to Collect Evidence:

Data Table and/or Results:

Evidence-based Conclusion Statement:

Phase VI: Summary

Create a 50-word summary, in your own words, that describes how the Sun's motion and position changes over the day and over the course of a year. Feel free to reference things learned in specific phases of this lab and to create and label sketches to illustrate your response.

3	Monitoring the Moving Constellations

Big Idea: Sky objects have properties, locations, and predictable patterns of movement that can be observed and described.

Goal: Students will conduct a series of inquiries about the motion of the constellations in the sky using sky simulation software and learn how different stars are visible at different times of the year in different locations in the sky.

Computer Setup:

This computer setup is the same as in the "Observing the Sun" lab activity. If you followed the setup instructions in that lab and saved the new default settings, you can skip this part of the setup. If not, set up your computer as follows:

Open the program *Stellarium* on your lab computer. If you are working outside of the lab, you can download and install *Stellarium* for free from http://www.stellarium.org/.

Hover your mouse over the bottom-left hand side of the screen until a menu appears. Select the top button, "Location Window." This will allow you to set your location to anywhere around the world, or any of the planets (or Sun) in our solar system. For this lab, we will set our location to "State College, PA." In the search box (the one with the magnifying glass next to it) in the location window, type "State College" and select the location from the list (it should be the only one). Click "Use as default" (if not already selected) so that the location is set to State College for any future uses. Simply close the location window to save these settings. *Be sure when you've set State College as your location that the arrow on the map is pointing to State College. If it is not, then you have selected the wrong location.*

The time is set to the current (system) time by default, so the sky you see on your screen should appear as it does right now outside (except for any weather). If you open the "Date/time" window, the time is shown in military time (i.e. 24 hour clock, so 6:00PM is represented as 18:00 and so on). We want to be able to see the constellations at any time of day, so hover over the left-hand menu and select "Sky and viewing options window." Uncheck the box that says "Show Atmosphere." Both removing the atmosphere and reducing light pollution will allow us to see the sky without any residual light getting in the way. While these are not realistic viewing conditions, they will let us gather more information for this activity.

Make sure that both the Constellation Boundaries and Constellation Labels are visible. The directions in this lab activity assume that you understand how to use the basic *Stellarium* settings of date, time, location, and constellation markers, as outlined in the lab activity, "Observing the Sun's Position and Motion."

Phase I: Exploration

1) When you first turn on *Stellarium*, the Sun is probably visible (if not, change the time until it is above the south-western horizon). If you were to go outside right now, could you see these stars shown on the simulation? *Explain why or why not.*

2) To which constellation is the Sun closest?

3) If you increase the time by one hour, remembering to use a 24-hour clock, toward which direction does the Sun move?   *Circle one:* North | South | East | West

4) Now, one hour later than when you started, to which constellation of stars is the Sun now closest?

5) If you advance the time to sunset, to which constellation of stars is the Sun closest at sunset?

6) Advance the time to sunrise tomorrow. To which constellation of stars is the Sun closest at sunrise tomorrow?

7) In a complete sentence, what general statement can you make about how the Sun and stars appear to move together through the sky over the course of the day (24 hours)?

Phase II: Does the Evidence Match the Conclusion?

Consider the research question, **"In which direction does the Sun move compared to the background constellations?"**

Set *Stellarium* to noon today. For the following days listed in the chart below, assuming that you could see the stars hidden behind the brilliantly shining Sun, find the constellation of stars closest to the Sun. *Note: while you're changing the date, make sure you're keeping the time the same! Check to make sure that you paused the rotation of the sky before collecting these data.*

Time from Today at Noon	Date	Constellation
0 days (i.e. today at noon)		
One day		
One week		
Two weeks		
Three weeks		
One Month		
Two Months		
Three Months		
Six Months		
Nine Months		
One Year		
Two Years		

8) If a student proposed a generalization that **"the constellations seem to slowly drift westward compared to the position of the Sun, with the Sun covering constellations at a rate of about one per week,"** would you agree or disagree with the generalization based on the evidence you collected by analyzing the pattern of how the Sun's position changes compared to the constellations? *Explain your reasoning and provide evidence either from the above questions or from evidence you yourself generate using* Stellarium.

Phase III: What Conclusions Can You Draw from the Evidence?

Orion is a prominent constellation visible in the winter time in the northern hemisphere, usually being hidden by the shining Sun in the summer. What conclusions and generalizations can you make from the following data collected that answers the question, **"When is Orion visible directly above the southern horizon as seen from State College?"** *Explain your reasoning and provide evidence to support your conclusion.*

Date	Time above Southern Horizon	Azimuth
October 1, 2013	05:07	180^0 (South)
November 1, 2013	03:05	180^0 (South)
December 1, 2013	01:07	180^0 (South)
January 1, 2014	23:02	180^0 (South)
February 1, 2014	21:00	180^0 (South)

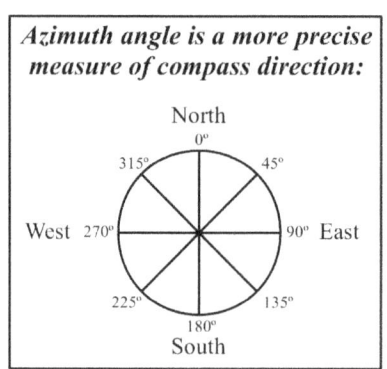

Azimuth angle is a more precise measure of compass direction:

9) Evidence-based conclusion:

Phase IV: What Evidence Do You Need?

Describe precisely what evidence you would need to collect in order to answer the research question, **"Over what precise period of time is my zodiac birth sign being covered by the Sun and is thus unable to be observed?"** You do not need to actually complete the steps in the procedure you are writing.

10) Create a detailed, step-by-step description of the evidence needed and a complete explanation of how this could be done using *Stellarium*. Your procedure should be more than just "look and see when the Sun is nearby," but exactly what someone completely unfamiliar with the program would need to do, step-by-step to accomplish this. This includes any non-default settings that are needed to collect the evidence as well as how to determine your horoscope birth sign, but *shouldn't* include meaningless filler like "analyze data and draw conclusions."

Phase V: Formulate a Question, Pursue Evidence, and Justify Your Conclusion

Your task is to design an answerable, original research question about some motion or position of the constellations in the sky that we have not previously addressed, propose a plan to pursue evidence, collect data using *Stellarium* by following your outlined procedure, and create an evidence-based conclusion that answers your research question.

Specific Research Question:

Step-by-Step Procedure to Collect Evidence:

Data Table and/or Results:

Evidence-based Conclusion Statement:

Phase VI: Summary

Create a 50-word summary, in your own words, that describes which constellations are visible at night and how this changes over the night and over the year. Feel free to create and label sketched to illustrate your response.

Your Name: _____

Group Member(s): _____

Big Idea: Weather is a snapshot description of Earth's atmosphere conditions at a particular location and at a particular time that is characterized by temperature, humidity, cloud cover, precipitation, barometric pressure, and wind speed.

Goal: To complete several scientific inquires about changing weather conditions at various locations.

Computer Setup and/or Materials Needed: Internet access to http://www.wunderground.com/history/

Phase I: Exploration

1) In the **Location** box, enter in your current location for today and complete the first blank column of the table below. Then, change the date to yesterday and then one year ago today and complete the remaining two columns.

Location:_____	TODAY	YESTERDAY	1 YEAR AGO TODAY
Maximum Temperature (using °F)			
Minimum Temperature (using °F)			
Average Humidity (using %)			
Day's Precipitation (using inches)			
Barometric (or Sea Level) Pressure*			
Wind Speed (mph)			

(*using inches Hg)

2) For YESTERDAY, make a few rough sketches of how the temperature, barometric pressure and wind speed average changed throughout that day. *Be sure to clearly label the vertical axes.*

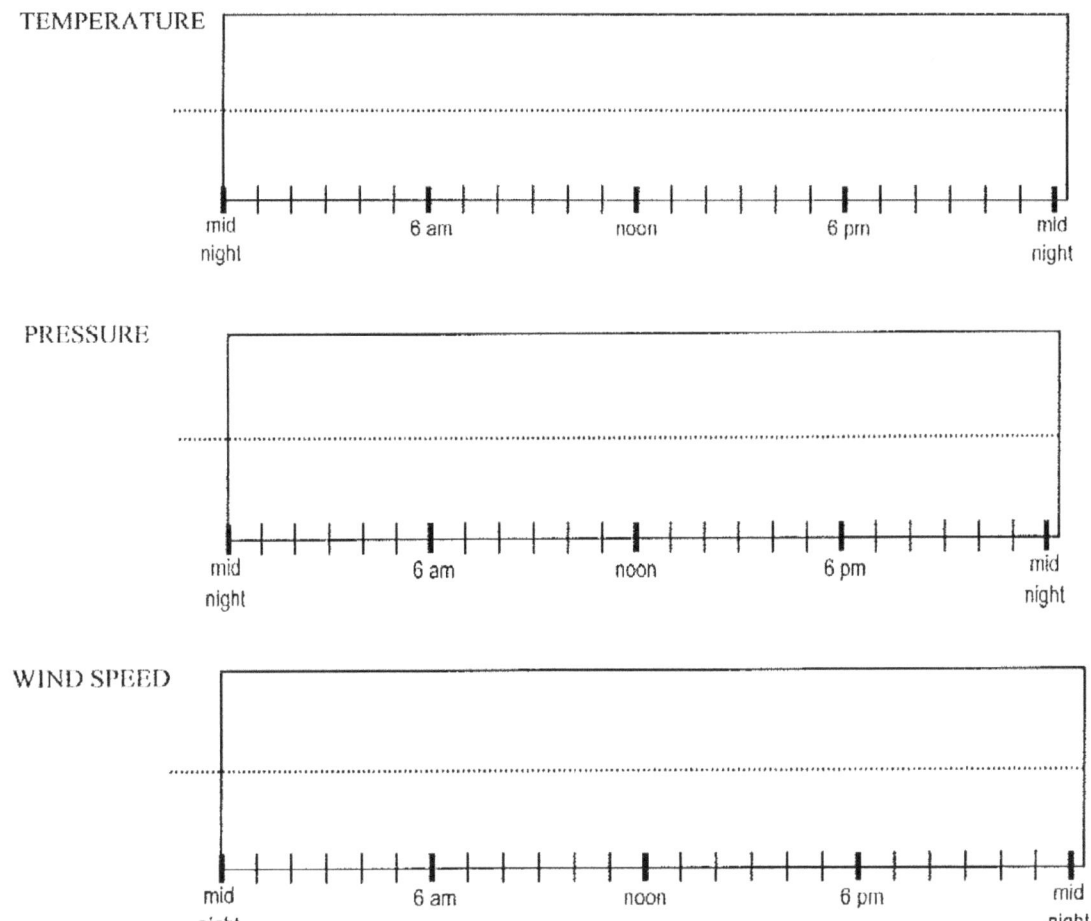

3) Consider the research question, "**How are the weather conditions here today different than yesterday?**" Write a paragraph comparing the weather yesterday and today. Be sure to include temperature, humidity, precipitation, pressure, and wind speed.

Phase II: Does the Evidence Match a Given Conclusion?

4) Consider the research question of "How much does the weather change year to year?" Now a student proposed a generalization that "**the temperature here today is about the same as it was at this same location on this same date, but back in the year you were born**." Would you agree or disagree with the generalization based on patterns you can find in the evidence you collected in the previous section or using new evidence? *Explain your reasoning and provide specific evidence either from the above questions or from any new evidence you yourself generate using this website.*

Phase III: What Conclusions Can You Draw from This Evidence?

Wind is caused when air rapidly moves from one place to another. What conclusions and generalizations can you make from the following data collected by a student to address the question **"At what time of year is it windiest in Laramie, Wyoming?"** by analyzing which season (winter, spring, summer or fall) has the greatest average wind speed. *Explain your reasoning and provide specific evidence from data, with sketches if necessary, to support your reasoning.*

Month (2008)	Average wind Speed (mph)	Wind Direction	Precipitation (inch)	Humidity (%)
Jan	21	WNW	0.01	62
Feb	4	WSW	0	82
Mar	9	South	0	62
Apr	22	SW	0	26
May	11	NNW	0.02	65
Jun	8	SSE	0	40
Jul	7	South	0	34
Aug	6	ENE	0.48	86
Sep	6	SSE	0	54
Oct	10	SSW	0	57
Nov	5	South	0	58
Dec	11	SSW	0	62

5) Evidence-based conclusion:

Phase IV: What Evidence Do You Need to Pursue?

Imagine your team has been assigned the task of designing a scientific observation plan for determining where to build windmills for electricity. Describe precisely what evidence you would need to collect and how to collect it in order to answer the research question of **"Where is it windiest in my state?"** You do not need to actually complete the steps in the procedure you are writing, but you might need to sketch a map of your state.

6) *Create a detailed, step-by-step description of evidence that need to be collected and a complete explanation of how this could be done- not just "look and see where it is windiest," but exactly what would someone need to do, step-by-step, to accomplish this. You might include a table and sketches – the goal is to be precise and detailed enough that someone else could follow your procedure.*

Phase V: Formulate a Question, Pursue Evidence, and Justify Your Conclusion

Your task is to design an answerable research question, propose a plan to pursue evidence, collect data and create an evidence-based conclusion about an aspect of weather that we have not previously addressed.

Specific research question:

Step-by-step procedure to collect evidence:

Data table and/or results:

Evidence-based conclusion statement:

Phase VI: Summary

Create a 50-word summary, in your own words, that describes weather patterns at different locations and how this changes over the year. Feel free to create and label sketches to illustrate your response.

5	Assessment Case Studies #1

Big Idea: Designing a fruitful plan for conducting research has many pitfalls. By assessing the research reports of others, scientists can improve their own ability to design attractive research plans. With better research designs, researchers can improve the support for the claims they make with better and better evidence.

Goal: Students will assess a series of research reports and then select one project to redesign and conduct in order to more productively pursue the original research question.

Assess Research Projects and Identify Inconsistencies in Their Lines of Inquiry:

In this lab you will identify any issues with research projects similar to those you have already completed. Your task is to distinguish between severe problems that could invalidate the research, and those that are minor. Work on improving only one research report at a time. Make sure to specify which report you are using by completely writing out the research question. Answer each of the questions by providing a short, but detailed, explanation of your reasoning citing specific information from the provided research reports.

Inquiry Research Report #11
Monitoring the Moving Constellations

Formulate a Question, Pursue Evidence, and Justify Your Conclusion

Your task is to design an answerable research question, propose a plan to pursue evidence, collect data using Stellarium (or another suitable source pre-approved by your lab instructor), and create an evidence-based conclusion about some motion or position in the sky for the constellations that you have not completed before.

Research report:

Specific research question:

During which season is the constellation Orion high in the southern sky just after sunset?

Step-by-step procedure to collect evidence:

Using Stellarium to make observations:
1. *Choose a day of the month to observe.*
2. *On each observation day, just after the sunset, determine if Orion is visible and determine in which part of the sky it is located.*
3. *Repeat the observation once a month for a year.*

Data table and/or results:

Date	Visible	Location		Date	Visible	Location
4/1/09	yes	high SW sky		10/1/09	no	n/a
5/1/09	yes	low W sky		11/1/09	no	n/a
6/1/09	no	n/a		12/1/09	no	n/a
7/1/09	no	n/a		1/1/10	yes	low E sky
8/1/09	no	n/a		2/1/10	yes	high SE sky
9/1/09	no	n/a		3/1/10	Yes	high S sky

Evidence-based conclusion statement:

From the evidence above, we can see that the constellation Orion appears to move from low in the eastern sky to low in the western sky from January to May.

Report analysis

Report number: _____

Keywords from Research question: _____

1) Write down some of the things you might observe to pursue this research question. *If you already took notes while reading the report, then there is no need to copy your notes here.*

2) List any problems with this report. **Make separate lists** for *minor* problems and *major* problems. *You may want to consider the following questions in determining whether something is a major or a minor problem. (You can use the space in the previous question to make notes.)*

   a) *Did they collect relevant evidence?*
   b) *Have they collected enough evidence? Or is their evidence insufficient and anecdotal?*
   c) *Did they claim more than the evidence supports?*
   d) *Did they follow their procedure?*
   e) *Do they answer the research question?*
   f) *Have assumptions impacted their results? That is, have the researchers made use of unjustified prior knowledge in lieu of collecting data?*

3) Do the major problems invalidate the research? *If no, explain why you classified them as major rather than minor.*

4) Do the minor problems invalidate the research? *If yes, explain why you classified them this way.*

5) Is the presentation of their results clear and unambiguous? What about the rest of the report? Anything that stands out as good or bad in their presentation of the report? *(If you already identified these problems as minor or major, just reference them here. No need to repeat yourself.)*

6) Precisely what should the researchers have done or reported differently to improve their research project?

Inquiry Research Report #12
Observing the Sun's Position and Motion

Formulate a Question, Pursue Evidence, and Justify Your Conclusion

Your task is to design an answerable research question, propose a plan to pursue evidence, collect data using Stellarium (or another suitable source pre-approved by your lab instructor), and create an evidence-based conclusion about some motion or position of the sun in the sky that you have not completed before.

Research report:

Specific research question:

Over the course of a year, how does the amount of Sunlight each day at the equator compare to that of Laramie, Wyoming?

Step-by-step procedure to collect evidence:

Use Stellarium to collect evidence:
 1. *Choose a number of days over the course of a year to make observations.*
 2. *Observe and record the Sunrise and sunset times on each of the observation days in Laramie, Wyoming and at the equator.*
 3. *Use the Sunrise and sunset times to calculate the total amount of daylight on each of the observation days and at both locations.*

Data table and/or results:

(All times are Mountain Standard Time)

Date	Laramie sunrise	Laramie sunset	Total daylight	Equator sunrise	Equator sunset	Total daylight
3/8/2009	6:25am	5:50pm	11:35	6:10am	6:08pm	11:58
3/15/2009	6:15am	6:00pm	11:45	6:09am	6:06pm	11:57
3/22/2009	6:00am	6:05pm	12:05	6:08am	6:04pm	11:56
3/29/2009	5:50am	6:15pm	12:25	6:05am	6:02pm	11:57

Evidence-based conclusion statement:

Over the course of a year, the Sun rises earlier and sets later in the summer than in winter in Laramie, Wyoming, so there is more daylight in summer and less in winter. The equator does not experience seasons so there is no change in the amount of daylight as the year progresses.

Report analysis

Report number: _____

Keywords from Research question: _____

7) Write down some of the things you might observe to pursue this research question. *If you already took notes while reading the report, then there is no need to copy your notes here.*

8) List any problems with this report. **Make separate lists** for *minor* problems and *major* problems. *You may want to consider the following questions in determining whether something is a major or a minor problem. (You can use the space in the previous question to make notes.)*

   a) *Did they collect relevant evidence?*
   b) *Have they collected enough evidence? Or is their evidence insufficient and anecdotal?*
   c) *Did they claim more than the evidence supports?*
   d) *Did they follow their procedure?*
   e) *Do they answer the research question?*
   f) *Have assumptions impacted their results? That is, have the researchers made use of unjustified prior knowledge in lieu of collecting data?*

9) Do the major problems invalidate the research? *If no, explain why you classified them as major rather than minor.*

10) Do the minor problems invalidate the research? *If yes, explain why you classified them this way.*

11) Is the presentation of their results clear and unambiguous? What about the rest of the report? Anything that stands out as good or bad in their presentation of the report? *(If you already identified these problems as minor or major, just reference them here. No need to repeat yourself.)*

12) Precisely what should the researchers have done or reported differently to improve their research project?

Inquiry Research Report #13
Monitoring the Zodiac Constellations

Formulate a Question, Pursue Evidence, and Justify Your Conclusion

Your task is to design an answerable research question, propose a plan to pursue evidence, collect data using Stellarium, and create an evidence-based conclusion about some motion or position in the sky for the constellations that you have not completed before.

Research report:

Specific research question:

Does the maximum altitude of the line of the zodiac constellations through the sky change over the course of a year in the same way as the path of the Sun? That is, does it move higher and higher above the southern horizon in spring and move lower and lower toward the southern horizon in fall?

Step-by-step procedure to collect evidence:

Use Stellarium to make observations:
  1. Choose a day of the month to observe the zodiac.
  2. Each month, right after sunset, record the direction and time the Zodiac is rising.
  3. For clarity, label directions as northeast, NE; east-northeast, ENE; east, E; east-southeast, ESE; etc.

Data table and/or results:

Date: Day of the year for observation; Time: Time after sunset when zodiac was observed; Direction: The direction on the horizon where the zodiac was rising

Date	Time	Direction		Date	Time	Direction
1/15/2009	5:00pm	ENE		7/15/2009	8:30pm	ESE
2/15/2009	5:30pm	ENE		8/15/2009	8:00pm	ESE
3/15/2009	7:15pm	E		9/15/2009	7:15pm	E
4/15/2009	7:45pm	E		10/15/2009	6:15pm	E
5/15/2009	8:15pm	ESE		11/15/2009	4:45pm	ENE
6/15/2009	8:45pm	ESE		12/15/2009	4:30pm	ENE

Evidence-based conclusion statement:

Every month the path of the zodiac appears to move lower and lower from summer all the way through fall. After winter starts, the path moves higher and higher above the horizon throughout spring until summer. This is nearly identical to the changes that the path of the Sun takes throughout the year, and also accounts for the changes in the seasons.

Report analysis

Report number: _____

Keywords from Research question: _____

13) Write down some of the things you might observe to pursue this research question. *If you already took notes while reading the report, then there is no need to copy your notes here.*

14) List any problems with this report. **Make separate lists** for *minor* problems and *major* problems. *You may want to consider the following questions in determining whether something is a major or a minor problem. (You can use the space in the previous question to make notes.)*

   a) *Did they collect relevant evidence?*
   b) *Have they collected enough evidence? Or is their evidence insufficient and anecdotal?*
   c) *Did they claim more than the evidence supports?*
   d) *Did they follow their procedure?*
   e) *Do they answer the research question?*
   f) *Have assumptions impacted their results? That is, have the researchers made use of unjustified prior knowledge in lieu of collecting data?*

15) Do the major problems invalidate the research? *If no, explain why you classified them as major rather than minor.*

16) Do the minor problems invalidate the research? *If yes, explain why you classified them this way.*

17) Is the presentation of their results clear and unambiguous? What about the rest of the report? Anything that stands out as good or bad in their presentation of the report? *(If you already identified these problems as minor or major, just reference them here. No need to repeat yourself.)*

18) Precisely what should the researchers have done or reported differently to improve their research project?

Choose One Research Project to Redesign, Improve, and Conduct

Your task is to choose one of the research projects (either report 12 or report 13) to redesign and carry out. You should re-use the <u>exact same research question as the previous researchers</u>, but make sure to improve the research design so that you eliminate the main problem(s) you were able to identify. Note that the next page contains an empty grid that may help you make any graphs more precise.

Your *redesigned* research report:

Specific research question:

Step-by-step procedure to collect evidence:

Data table and/or results:

Evidence-based conclusion statement:

19) Precisely what has been done or reported differently to improve the original research inquiry project?

Summary

20) Write a 50-word summary of what makes a solid inquiry research project. Explain what the biggest problems were, and how you corrected them. Be sure to describe details about how your changes improved the line of inquiry.

21) Are numbers useful?

22) Are graphs useful?

6   Observing Jupiter's Moons

Big Idea: Sky objects have properties, locations, and predictable patterns of movements that can be observed and described.

Goal: Students will conduct a series of inquiries about the position and motion of Jupiter's moons using prescribed Internet simulations.

Computer Setup:

Access http://space.jpl.nasa.gov/ and

a) Select THE EARTH in the "Show me_____" drop-down menu.

b) Select THE SUN in the "as seen from" drop-down menu.

c) Select the radio button "I want a field of view of __ degrees," and set the drop-down menu to 0.5.

d) Select the check box for EXTRA BRIGHTNESS and then select "Run Simulator."

Phase I: Exploration

1) The resulting image shows what one would see looking through a special telescope. In this picture, where is the observer with the special telescope located?

2) How does the image change if you INCREASE the field of view?

3) What is the exact date of the image?

4) Astronomers typically mark images based on the time it currently is in Greenwich, England, called UTC. What is the precise time listed on the image?

5) Use a ruler to measure the distance on the screen between the center of Earth and the center of the Moon. Record in the table the measured Earth-Moon distance and location (left or right) at the times listed below. You do NOT need to know the exact number of kilometers, but simply a ruler-measurement you can compare to other measurements you make later. You can use the Squidgit ruler found on the last page of this lab, or you can use the edge of a blank piece of paper held in the landscape orientation to mark the distance between the Earth and Moon.

Time from when you started	Earth-Moon Distance	Direction (left or right)
0 (i.e. when you started)		
One Hour		
One Day		
Three Days		
Five Days		
Ten Days		
Two Weeks		
One Month		
Three Months		

6) Consider the research question of, **"How long does it take the Moon to orbit Earth?"** The typical time that it takes the Moon to make a full orbit the Earth is one month. Which of your observations confirms or contradicts this statement? *Explain, and consider including a sketch of the orbit as viewed from above.*

Phase II: Does the Evidence Match the Conclusion?

7) Consider the research question, **"How long does it take one of Jupiter's moons to orbit Jupiter?"** Set the Solar System Simulator to observe Jupiter from the Sun, where Jupiter takes up 10% of the image. Measure the distance between Jupiter and Io shown on the image at the times listed below.

Time from when you started	Jupiter-Io distance	Direction (left or right)
0 (i.e. when you started)		
0.5 days		
1.0 days		
1.5 days		
2.0 days		
2.5 days		
3.0 days		
3.5 days		
4.0 days		
4.5 days		
5.0 days		
5.5 days		
6.0 days		

8) If a student proposed a generalization that **"Io orbits the planet Jupiter about once every 48 hours,"** by noting patterns in the time it takes for Io to return to its original position from where it started, would you agree or disagree with the generalization based on the evidence you collected? *Explain your reasoning and provide specific evidence either from the above questions or from evidence you yourself generate using the Solar System Simulator.*

Phase III: What Conclusions Can You Draw from the Evidence?

Europa is one of the four largest moons orbiting Jupiter. The others are Io, Callisto, and Ganymede. What conclusions and generalizations can you make from the following data collected by a student in terms of the question **"How long does it take Europa to orbit Jupiter?"** *Explain your reasoning and provide specific evidence to support your reasoning. Consider including a sketch of the orbit as viewed from above.*

Time	Measured Distance from Jupiter	Appearance Notes
11 PM Monday	0 squidgits	Not visible, likely behind Jupiter
11 PM Tuesday	5.0 squidgits	On Jupiter's left side
11 PM Wednesday	1.5 squidgits	On Jupiter's right side
11 PM Thursday	5.0 squidgits	On Jupiter's right side
11 PM Friday	No observations	Cloudy

9) Evidence-based conclusion:

Phase IV: What Evidence Do You Need?

Imagine your team has been assigned the task of writing a news brief for your favorite news blog about the length of time it takes Ganymede, the largest moon in the entire solar system, to orbit Jupiter once. Describe precisely what evidence you would need to collect in order to answer the research question of, **"Over what precise period of time does it take Ganymede to orbit Jupiter?"** You do not need to complete the procedure you've written.

10) *Create a detailed, step-by-step description of evidence that needs to be collected and a complete explanation of how this could be done—not just "look and see when Ganymede is first on one side and then on the other," but exactly what would someone need to do, step-by-step, to accomplish this. You might include a table and sketches—the goal is to be precise and detailed enough that someone else could follow your procedure. Do not include meaningless filler like "Find evidence" or "Analyze data and draw conclusions."*

Phase V: Formulate a Question, Pursue Evidence, and Justify Your Conclusion

Your task is to design an answerable research question, propose a plan to pursue evidence, collect data using Solar System Simulator (or another suitable source pre-approved by your lab instructor), and create an evidence-based conclusion about some motion or changing position of a moon of the solar system that we have not previously addressed.

Specific research question:

Step-by-step procedure, with sketches if needed, to collect evidence:

Data table and/or results:

Evidence-based conclusion statement:

Phase VI: Summary

Create a 50-word summary, in your own words, that describes the motions, orbits, or rotations of Jupiter's moons and how this changes over time. You should cite specific evidence you have collected in your description, not describe what you have learned in class or elsewhere. Feel free to reference things learned in specific phases of this lab, or to create and label sketches to illustrate your response.

Astronomical Ruler

(units of squigits)

	1
	2
	3
	4
	5
	6
	7
	8
	9
	10
	11
	12
	13
	14
	15
	16
	17
	18
	19
	20
	21
	22
	23
	24
	25
	26
	27
	28
	29
	30

| 7 | **Studying Exoplanets** |

Big Idea: Planets orbiting other stars (exoplanets) have orbital characteristics similar and different to our own solar system of planets orbiting our Sun.

Goal: Students will conduct a structured series of scaffolded scientific inquiries about the nature of observed exoplanets using the Internet site prescribed, particularly the *Exoplanet Data Explorer*.

Computer Setup: Access URL http://exoplanets.org/

Solar System Data Table

Name	Planet Mass (M$_{Earth}$)	Planet Mass (M$_{Jupiter}$)	Orbital Period (Earth Years)	Semi-Major Axis (AU)
Mercury	0.06	0.0002	0.24	0.39
Venus	0.82	0.003	0.62	0.72
Earth	1.00	0.003	1.00	1.00
Mars	0.11	0.0003	1.88	1.52
Jupiter	318	1.00	11.86	5.20
Saturn	95.2	0.299	29.5	9.54
Uranus	14.5	0.046	84.0	19.2
Neptune	17.1	0.054	165	30.1

The table above lists basic information about the eight planets of our Solar System, listed in order of their nearness to the Sun. Planet mass has been listed in units of both M$_{Earth}$ (spoken as "Earth Masses") and M$_{Jupiter}$ (spoken as "Jupiter Masses"), as these are the two mass units that are most often used when describing exoplanets. These units describe the mass of a planet in comparison to either Earth or Jupiter (i.e. Earth's mass is 1 M$_{Earth}$ and Jupiter's mass is 1 M$_{Jupiter}$). Orbital period and planet semi-major axis (orbital distance) are also listed in the most common units used to describe exoplanets. As a reminder, "AU" stands for "Astronomical Unit," which is the average distance from the Earth to the Sun.

Phase I: Exploration

This section concerns the planets in our own solar system; Phase II and beyond concerns exoplanets.

1) A **histogram** is a bar-chart showing the number of objects in a particular category, so it is useful for showing how populations are distributed in a certain characteristic. Consider the research question, "How are characteristics of planets distributed?" Use the Solar System Data Table and sketch the three histograms described below.

Histogram 1: Distribution of Orbital Distance. Make a histogram showing the number of Planets closer and farther than or equal to Earth's orbital distance, using the axes given below.

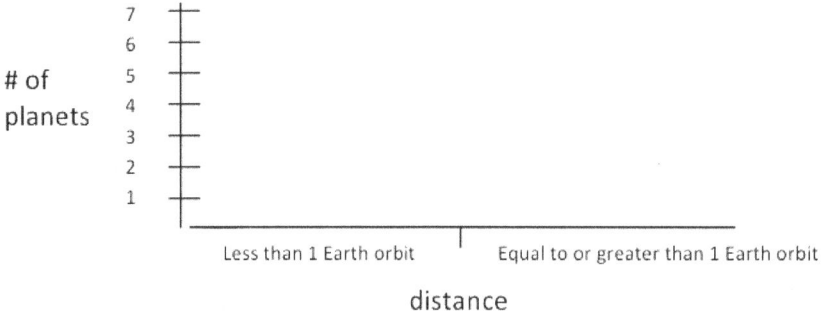

Histogram 2: Distribution of Masses. Make a histogram showing the number of planets with masses less than Earth's mass and greater than or equal to Earth's mass, using the axes given below.

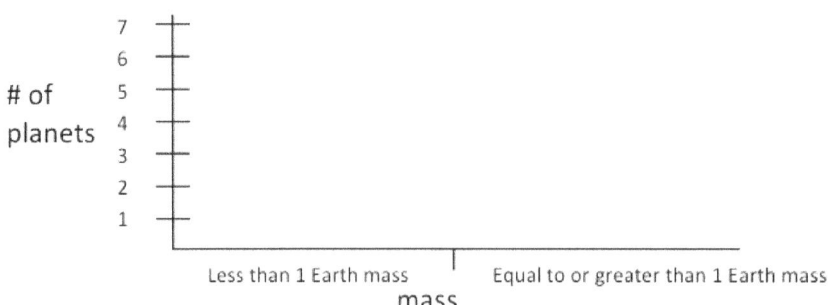

<u>Histogram 3: Distribution of Orbital Periods.</u> Make a histogram showing the number of planets with orbital periods smaller than Earth's period (P < P_{Earth}), between Earth's and Jupiter's periods ($P_{Earth} \leq$ PERIOD $\leq P_{Jupiter}$), and longer than Jupiter's period (P > $P_{Jupiter}$), using the axes given below. Note: Throughout this lab, "Period" and "Orbital Period" are synonymous.

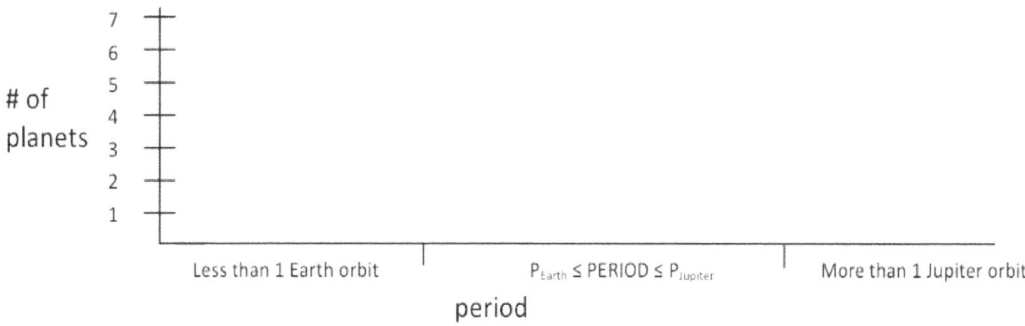

2) Consider the research question, "What is the distribution of orbital distances for exoplanets?" A correlation-diagram (or "scatter plot") is a graph of dots showing how two characteristics, or variables, are related. Use the Solar System Data Table and sketch a correlation-diagram (graph) for each of the following descriptions using the corresponding axes. Make sure to label your axes with the appropriate characteristic name, the units used, and tick marks labeled with numbers to show the scale of the graph.

<u>Scatter Plot 1:</u> Distance (AU) vs. Period (Years) for Planets Closer than Jupiter (*not including Jupiter*). (Vertical Y-axis Distance versus Horizontal X-axis Period).

<u>Scatter Plot 2:</u> Distance (AU) vs. Period (Years) for Planets With Orbits Jupiter-sized and larger. (Vertical Y-axis Distance versus Horizontal X-axis Period).

<u>Scatter Plot 3:</u> Distance (AU) versus Mass (M_{Earth}, which means in units of Earth's mass) for ALL Solar System Planets. (Vertical Y-axis Distance versus Horizontal X-axis Mass).

3) Consider the research question, "Which characteristics of exoplanets are most highly correlated with distance?" The notion of **correlation** is the idea that two characteristics are closely related to one another. *IMPORTANT NOTE: CORRELATION IS NOT THE SAME AS CAUSE-AND-EFFECT.*

To illustrate the notion of correlation, let us consider the two graphs next. One of the two graphs is *Intelligence versus Height* and the other is *Weight versus Height*.

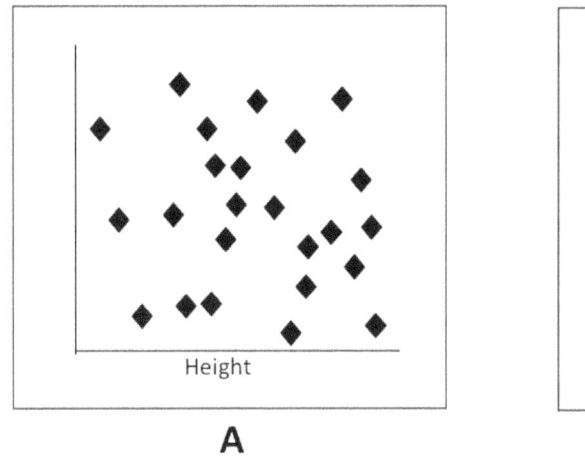

A

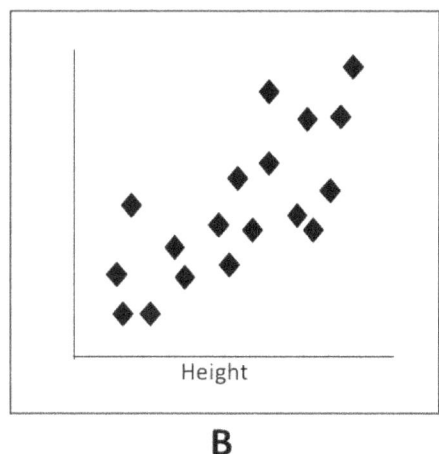

B

Which graph represents Weight vs. Height and which graph represents Intelligence vs. Height? Explain how you know this.

4) Based on your work above on analyzing the planets of our solar system, which variable, PERIOD or MASS, seems to be more highly *correlated* to DISTANCE? Explain your reasoning, using any needed labeled sketches, if you like, to illustrate your answer.

Phase II: Does the Evidence Match a Given Conclusion?

A Brief Tutorial On Using The Exoplanets Data Explorer Table at http://exoplanets.org/table

a. Hovering your mouse over a column header shows an explanation of what each term means. This is true later on, using the "Plot" feature as well.

b. Sorting: Clicking on a column header sorts the data table by that quantity. Clicking it again reverses the order of the sort.

c. Notice that the first column gives the exoplanet's NAME. *Try sorting the table by NAME.*

d. The second column is the exoplanet MASS (times a factor called "sin(i)", which we will ignore because it has little effect). The MASS of the planet is given in terms of how many times bigger (or smaller) than the mass of our planet Jupiter, $M_{jupiter}$ by default, but you can change the units by clicking on this label and selecting from a drop-down list of alternate units. *Try changing the units of mass from ' $M_{jupiter}$ ' (Jupiter Mass) to ' M_{Earth} ' (Earth Mass), and then back to Jupiter Mass again.*

e. The third column shows the SEMI-MAJOR AXIS. This is another name for how far the planet orbits its star, on average. The default units of distance are AU, or Astronomical Unit. IMPORTANT DEFINITION: One AU is the average distance our Earth orbits our Sun.

f. The fourth column shows the exoplanet's Orbital Period, a.k.a. PERIOD. The period is the length of time it takes the planet to go around its central host star once. By default the units are Earth days, but you can change the units by clicking on this label. *Try sorting the table from largest to smallest period.*

g. Removing Columns: You can simplify the table by removing columns you don't want to look at. If you hover your mouse over a column header, you should see a faint red "x" that allows you to remove that column. *Try this with "Time of Periastron" as an example.* You can always add a column back in after removing it.

h. Adding Columns: You can add columns to the table by clicking the large "+" (plus) sign at the top right of the page. There are many categories to choose from! *Add in the column "Date", as we will need it.*

5) PART A: Access the Exoplanet Data Explorer [http://exoplanets.org/], **"Table"** option, and sort and search the data to find a planet that was discovered (published) in 1995 and record data about it here. You will find it helpful to add a First Publication "Date" column to the table! (See the mini-tutorial above for instructions for adding columns.) The units are provided for Mass, but you must fill them in for Period and Semi-Major Axis.

Planet Name: _____		
Property	**Value**	**Units**
Mass		Jupiter Masses
Period		
Semi-major Axis		

6) Is this planet more massive than Earth?   Yes   No

7) If Yes, how many more times more massive? If No, what fraction of Earth's mass does it have?

PART B: Select "Plots" at the top left, then "Histogram Plot" at the right. Choose *Semi-Major Axis* as the "Data" to plot. (It's in the third subsection, under "Orbit Parameters".) All confirmed planets to date will be shown by default. *Remember that Earth orbits our Sun at a distance of 1 AU and Jupiter orbits at about 5 AU.*

8) Click "Add Filter" to see the number of planets (#) under the *Statistics After Cut* section (with no filter criteria selected, this number reflects the total number). How many exoplanets are initially shown in this data set? (Note: the answer you get will depend on the day you do it, as this number is continually updated to reflect the current total.)

9) Clicking "Add Filter" lets you add a criterion to restrict the number of planets appearing on the plot. Under the "+" sign next to the "Filter" text box, choose "Semi-Major Axis". "A[au]" should now appear in the box. "A" is the abbreviation for semi-major axis, and "AU" are the units. To the right of this, in the box, if you type ">10", this will restrict the sample to planets whose orbits are larger than 10 AU. Notice that the number of planets is now very small, since there are few currently known exoplanets with orbits that large. If you instead change this to "<10", you will see something near the original number of planets back, because our instruments are most sensitive in detecting planets at these orbital distances.

How many of the currently known exoplanets have orbits larger than Jupiter's orbit about our Sun? _____

10) What is the *percentage* of currently known exoplanets that have orbits larger than Jupiter's orbit about our Sun? Your answer should be a number only between 0 and 100:
_____ %

11) How many of the currently known exoplanets have orbits smaller than Earth's orbit about our Sun?

12_ What is the *percentage* of currently known exoplanets that have orbits smaller than Earth's orbit about our Sun? Your answer should be a number only between 0 and 100:
_____ %

PART C: Click the red "X" next to your filter to remove it. Still using "Histogram Plot", now choose *Orbital Period* as the "Data" to plot. All confirmed planets to date will be shown by default. *Remember that Earth orbits our Sun once every 365 days and Jupiter orbits once about every 4,300 days.*

13) How many exoplanets in total are shown in this particular data set? (Again, the exact numbers you get will depend on the day you do it)

14) What **percentage** of the planets shown has orbital periods similar to our planet Mercury? Say, <100 days? *Your answer should be a number between 0 and 100:* _____ %

15) What **percentage** of the planets shown has orbital periods similar to our planet Earth? Say, between 300 and 500 days? _____ %

16) Consider the research question, "How long do exoplanets take to orbit their star?". If a fellow student proposed a generalization that **"Most exoplanets discovered take about the same length of time to orbit their star as Earth takes to orbit our Sun,"** would you agree or disagree with the generalization based on the evidence you collected by looking at the range of possible orbital periods? *Explain your reasoning and describe specific evidence, with sketches if necessary, either from the above tasks or from new evidence you yourself generate using the Exoplanets Data Explorer.*

Phase III: What Conclusions Can You Draw from This Evidence?

What conclusions and generalizations can you make from the data organized using a <u>correlation diagram</u> (a.k.a. "scatter plot") in terms of the question **"how does the size of an exoplanet's orbit compare to its orbital period?"** *Explain your reasoning and provide specific evidence, with sketches if necessary, to support your reasoning. Remember, a picture is worth 10^3 words!* ***Optional:*** *Feel free to create and label sketches or graphs to illustrate your response. If available to you, you may export and print a scatter plot created with the web tool, and attach it to this lab writeup. Do not email your file to the instructor.*

<u>Collect Evidence:</u> Select "Scatter Plot" and choose the horizontal X-axis to be *Semi-Major Axis* (i.e. size of orbit) and the vertical Y-axis to be *Orbital Period* (i.e. time to complete an orbit). Expand the "Configure Axes" option at the top and try unchecking the "Log" boxes next to both X and Y, which makes the axes linear instead of logarithmic. (You should experiment with both types of axes in any plots that you make. Logarithmic scaling is often better at visually displaying data that are crowded or that cover a large range of values.) Once you have made a scatter plot, you can click and drag the graph around to center on different parts of it. You can zoom in or out on any portion of it by placing your mouse cursor over it and scrolling up or down. If your mouse doesn't have a scroll wheel, you can always set a Min and Max by hand under Axes Configuration.

20) Evidence-based conclusion:

Phase IV: What Evidence Do You Need?

Imagine your team has been assigned the task of predicting how far a newly discovered exoplanet would orbit from its central star. Describe precisely what evidence you would need to collect in order to answer the research question of, **"If an exoplanet were discovered to have an orbital period of 21 days, what would you predict its semi-major axis orbital distance to be using a correlation diagram/scatter plot?"** (This time the orbital period is the "independent," or X-axis variable, and the semi-major axis of the planet's orbit would be the "dependent," or Y-axis variable.) You do not need to actually complete the steps in the procedure you are writing.

21) *Create a detailed, step-by-step description of evidence that needs to be collected and a complete explanation of how this could be done - not just "look and see what value it would have", but exactly what would someone need to do, step-by-step, to accomplish this. You might include a table and sketches - the goal is to be precise and detailed enough that someone else could follow your procedure. Do NOT include generic nonspecific steps such as "analyze data" or "present conclusions" -- these are meaningless filler. Be specific!*

Remember, a picture is worth 10^3 words! **Optional:** *Feel free to create and label sketches or graphs to illustrate your response. If available to you, you may export and print a scatter plot created with the web tool, and attach it to this lab writeup. Do not email your file to the instructor.*

Phase V: Formulate a Question, Pursue Evidence, and Justify Your Conclusion

Your task is to design an answerable research question, propose a plan to pursue evidence, collect data using the Exoplanets Data Explorer (or another suitable source pre-approved by your instructor), and create an evidence-based conclusion about the characteristics of known exoplanets that we have not yet addressed. **REQUIRED this time:** Create and label sketches or graphs to illustrate your response. The Exoplanets Data Explorer has an "Export" button at the top right that will allow you to download your graphs.

Specific Research Question:

Step-by-Step Procedure to Collect Evidence:

Data Table and/or Results (If you include a graph from the web tool as part of your data, please attach it after this page):

Evidence-based Conclusion Statement:

Phase VI: Summary

Create a 50-word summary, in your own words, that describes the nature, frequency, or discovery of exoplanets and systems we have discovered so far. You should cite what you learned from doing *each* of the phases of this lab, not describe what you have learned in your lecture class or elsewhere.

8	Observing Features on the Sun

Big Idea: The Sun has surface features in different wavelengths and those features have predictable patterns of movement that can be observed and described.

Goal: Students will conduct a series of inquiries about the nature and motion of features in different wavelengths on the Sun using prescribed Internet observations that access actual solar images taken by satellites.

Computer Setup:
Access the SOHO Helioviewer at http://helioviewer.org/

1.a) Read about the Solar and Heliospheric Observatory (SOHO) in the following link:
http://sohowww.nascom.nasa.gov/about/docs/SOHO_Fact_Sheet.pdf

You only need to read the first page and a half, but you can read the rest if you're interested.

To see how active regions look at different temperatures, watch the following movie:
http://sohowww.nascom.nasa.gov/gallery/Movies/spotchange/spotchange.mov

Then, watch two or three movies of your choice from http://sohowww.nascom.nasa.gov/gallery/
Which two movies did you watch?

1.b) You will look at images in two wavelengths of light, 94 Angstroms (Å) and 4500 Å. What parts of the electromagnetic spectrum are these? (Radio, Microwave, Infrared, Visible, Ultraviolet, X-ray, Gamma Ray)

94 Å: _____   4500 Å: _____

With your mouse over the image of the Sun, click "m" on your keyboard. You should see two numbers on the bottom right of the image. These numbers measure your mouse's horizontal and vertical position on the image in *arcseconds*, measured from the center of the image. There are 60 *arcseconds* (") in one *arcminute* ('), and 60' in one degree(°).

1.c) **Look up:** How many times bigger is the diameter of the Sun than the diameter of Earth?

Measure: What is the diameter of the Sun in arcseconds?

Calculate: How big would the Earth be in the Helioviewer image, in arcseconds?

Phase I: Exploration

2) Click "Add" above the gray bar on the left of the screen, so there are two gray bars. Click the grey bars to display image options. For the images, use these settings:

Setting	Image 1	Image 2
Opacity	drag marker to right	drag marker to left
Observatory	SDO	SDO
Instrument	AIA	AIA
Detector	AIA	AIA
Measurement	4500 Å (visible)	94 Å (X-ray)

Keep the "Opacity" marker on the first image where it is. If you drag the "Opacity" marker on the second image, you'll notice that you can see either just the 4500 Å image or just the 94 Å image (Do you understand which image is which?). You're not really switching between images; you are just viewing the Sun through different wavelength filters as you move the "Opacity" marker from right to left. Moving the Opacity to the left makes the image fainter (all the way left makes it invisible).

3) Set the date to a week ago today, and adjust your settings so that the X-ray image is shown. Identify a feature (active X-ray region) and then increase the date by one day. How did the feature you identified in the earlier image move?

Circle one: left | right | up | down

4) Select the X-ray image and increase the date by one more day. Did the active X-ray region you identified in the earlier images appear to shift the same distance in the same direction as it did before?

Circle one: yes | no

5) Consider the research question, "Is there a relationship between the position of Sunspots in the visible light image and the X-ray image?" Find your feature on the X-ray image and then open the visible image for that same day. Is there clear evidence for a relationship between the precise position of the Sunspots and the active X-ray regions?

Circle one: yes | no

6) Create sketches in the space below to illustrate the conclusion you made in question 5, regarding whether or not features in X-ray images of the Sun correlate with features in visible images.

Phase II: Does the Evidence Match the Conclusion?

7) Choose a recent day for which the 4500 Å image has several Sunspots on it.

8) Use your mouse to measure and record the width of a few Sunspots in arcseconds (you may need to zoom in). What is the average size of Sunspots? How does this compare to the diameter of the Sun (in arcseconds) that we measured in question 1.c?

9) Consider the research question, "How large are Sunspots, on average?" If a student proposed a generalization that "**An average Sunspot is approximately the width of 10 Earth diameters,**" would you agree or disagree with that generalization based on the evidence you've collected or new evidence you need to collect? *Explain your reasoning and provide specific evidence either from the above questions or from evidence you yourself generate using the online data.*

Phase III – What Conclusions Can You Draw From the Evidence?

Imagine a solar scientist decided to look for evidence showing a relationship between the area of visible Sunspot activity and the area of X-ray activity, resulting in the following data table. The program recorded the size in square arcseconds of each area measured. What conclusions and generalizations can you make from the following data collected by a solar scientist related to the research question, **"How large are X-ray features compared to white light features?"** *Explain your reasoning and provide specific numeric evidence, with sketches if necessary, to support your reasoning.*

Date	X-ray [Square arcseconds]	Visible Light [Square arcseconds]
01/06/92	347	52
01/09/92	380	64
01/12/92	183	21
01/15/92	83	11
01/17/92	150	10

10) Evidence-based conclusion:

Phase IV: What Evidence Do You Need?

Describe precisely what evidence you would need to collect and how you would collect it in order to answer the research questions of, "**Over what precise period of time does it take an active region in the X-ray image near the Sun's equator to complete one full rotation?**"

11) *Create a detailed, step-by-step description of evidence that needs to be collected and a complete explanation of how this could be done—not just "look and see when the region returns to the same point," but exactly what would someone need to do, step-by-step, to accomplish this. You might include a table and sketches - the goal is to be precise and detailed enough that someone else could follow your procedure.*

Phase V: Formulate a Question, Pursue Evidence, and Justify Your Conclusion

Your task is to design an answerable research question, propose a plan to pursue evidence, collect data using *HelioViewer* (or another suitable source pre-approved by your lab instructor), and create an evidence-based conclusion about some relationships between features seen in images of different wavelengths or changing positions of a solar feature that we have not yet addressed.

Research Question:

Step-by-step procedure to collect evidence:

Data table, analysis, and/or results:

Evidence-based conclusion statement:

Phase VI: Summary

Create a 50-word summary, in your own words, that describes the features in different wavelengths on the Sun and how the motions or shapes of these features change over time. You should cite specific evidence you have collected in your description, not describe what you have learned in class or elsewhere. Feel free to reference things learned in specific phases of this lab and create and label sketches to illustrate your response.

| 9 | **Exploring GalaxyZoo – One** |

Big Idea: The countless galaxies of stars spread throughout the Universe have characteristics that can be observed and classified.

Goal: Students will conduct a structured series of scaffolded scientific inquiries about the nature of observed galaxies using the Internet sites prescribed, particularly the *Sloan Digital Sky Survey* via the original *Galaxy Zoo*.

Computer Setup:

Access http://zoo1.galaxyzoo.org/ and

a) Select REGISTER and set up a username and a password that is identical to your username. *Record this information on this sheet.*

b) Use one of your team member's email addresses. *Record it in the box.*

> Userid: _____
> Password: *same as above*
>
> Email: _____
>
> Security Question: **What is your favorite class?**
>
> Answer: **Astro11**

c) Create and enter a security question and answer. *Record it in the box.*

d) Select CREATE USER and, when set up, select CONTINUE.

Phase I: Exploration

1. From your book, your class notes, or from the Internet, find a reasonably good definition of a galaxy and write it in the space below (be sure to cite your source). *Include one rough sketch of what one looks like.*

    Definition and sketch of GALAXY:

2. What kind of astronomical object is the Milky Way? _____

3. What is the diameter of the Milky Way in miles? _____

4. How many stars are in the Milky Way? _____

5. How far away is the Andromeda Galaxy (M31), the nearest large galaxy? _____

6. Imagine that each of the two images below show a single galaxy of billions of stars.

Spiral Galaxy

Elliptical Galaxy

In the space below, create a detailed listing of the characteristics of each that allow you to distinguish one galaxy from the other.

Observable Characteristics of a Spiral Galaxy	Observable Characteristics of an Elliptical Galaxy

After logging into http://zoo1.galaxyzoo.org/, if you aren't already in, select HOW TO TAKE PART and complete the online TUTORIAL, completing the table below as you go along.

For each tutorial on an observable characteristic, first record your answers in the blocks above corresponding to the appropriate Tutorial Part. Then, check your answers and in the event your answer does not agree, mark a single line through your response. *There is no penalty for having an incorrect answer.*

7. TUTORIAL PART 1A: SPIRAL or ELLIPTICAL

8. TUTORIAL PART 1B: SPIRAL or ELLIPTICAL

9. TUTORIAL PART 1C: MERGING or NOT MERGING

10. TUTORIAL PART 1D: SPIRAL or ELLIPTICAL

11. TUTORIAL PART 2B: CLOCKwise or ANTI-clockwise or EDGE on/can't tell

12. TUTORIAL PART 3: GALAXY or "?"

Now you are ready to move forward!

13. At the bottom of the HOW TO TAKE PART – TUTORIAL page, select the PROCEED TO THE TRIAL button. Your team will be asked to fully classify 15 galaxies. If your classification agrees with scientists' classifications 8 or more times, you will be able to enter the *GalaxyZoo* scientific database to conduct your research. *You can repeat this step if necessary.*

<u>Record the results of your TRIAL data collection here:</u>

	Circle One		
Image #1	Clockwise Spiral	Anticlockwise Spiral	Edge On / Unclear
	Elliptical Galaxy		
	Star / Don't Know	Merging Galaxies	
Image #2	Clockwise Spiral	Anticlockwise Spiral	Edge On / Unclear
	Elliptical Galaxy		
	Star / Don't Know	Merging Galaxies	
Image #3	Clockwise Spiral	Anticlockwise Spiral	Edge On / Unclear
	Elliptical Galaxy		
	Star / Don't Know	Merging Galaxies	
Image #4	Clockwise Spiral	Anticlockwise Spiral	Edge On / Unclear
	Elliptical Galaxy		
	Star / Don't Know	Merging Galaxies	
Image #5	Clockwise Spiral	Anticlockwise Spiral	Edge On / Unclear
	Elliptical Galaxy		
	Star / Don't Know	Merging Galaxies	
Image #6	Clockwise Spiral	Anticlockwise Spiral	Edge On / Unclear
	Elliptical Galaxy		
	Star / Don't Know	Merging Galaxies	
Image #7	Clockwise Spiral	Anticlockwise Spiral	Edge On / Unclear
	Elliptical Galaxy		
	Star / Don't Know	Merging Galaxies	
Image #8	Clockwise Spiral	Anticlockwise Spiral	Edge On / Unclear
	Elliptical Galaxy		
	Star / Don't Know	Merging Galaxies	
Image #9	Clockwise Spiral	Anticlockwise Spiral	Edge On / Unclear
	Elliptical Galaxy		
	Star / Don't Know	Merging Galaxies	
Image #10	Clockwise Spiral	Anticlockwise Spiral	Edge On / Unclear
	Elliptical Galaxy		
	Star / Don't Know	Merging Galaxies	

Image #11	Clockwise Spiral	Anticlockwise Spiral	Edge On / Unclear
	Elliptical Galaxy		
	Star / Don't Know	Merging Galaxies	

Image #12	Clockwise Spiral	Anticlockwise Spiral	Edge On / Unclear
	Elliptical Galaxy		
	Star / Don't Know	Merging Galaxies	

Image #13	Clockwise Spiral	Anticlockwise Spiral	Edge On / Unclear
	Elliptical Galaxy		
	Star / Don't Know	Merging Galaxies	

Image #14	Clockwise Spiral	Anticlockwise Spiral	Edge On / Unclear
	Elliptical Galaxy		
	Star / Don't Know	Merging Galaxies	

Image #15	Clockwise Spiral	Anticlockwise Spiral	Edge On / Unclear
	Elliptical Galaxy		
	Star / Don't Know	Merging Galaxies	

14. Rate the relative DIFFICULTY your team has distinguishing the following by circling one on each line.

Rate the difficulty of classifying each of the following: Circle one	Nearly Impossible	Challenging	Some easy, some not	Pretty Easy
Elliptical Galaxies	Imposs.	Chall.	Varies	Easy
Spiral Galaxies	Imposs.	Chall.	Varies	Easy
Edge-on Spiral Galaxies	Imposs.	Chall.	Varies	Easy
Merging Galaxies	Imposs.	Chall.	Varies	Easy
Stars	Imposs.	Chall.	Varies	Easy
Cosmic Rays or Satellite Streaks	Imposs.	Chall.	Varies	Easy

Phase II: Does the Evidence Match the Conclusion?

Enter the GALAXY ANALYSIS Galaxy Zoo scientific database and classify ten (10) images. Keep a record of your results in the "Tally Sheet" below using tick marks. 𝗧𝗛𝗟

TALLY SHEET Data Table #1	Clockwise Spiral	Anticlockwise Spiral	Edge On / Unclear
	Elliptical Galaxy		
	Star / Don't Know	Merging Galaxies	

15. Consider the research question, "what type of galaxy is most common?" If a student proposed a generalization that "**most galaxies are elliptical**," would you agree or disagree with the generalization based on the evidence you collected SO FAR? Analyze the evidence of how many of each type of galaxy show up in your data tables to pursue this question. *Explain your reasoning and provide specific evidence either from the above questions or from evidence you yourself generate using* GalaxyZoo.

Phase III: What Conclusions Can You Draw From the Evidence?

Galaxies are observed to have numerous different shapes. Consider the research question, "which direction do spiral galaxies usually spin?" What conclusions and generalizations can you make from the following data collected by a group of students in terms of the question, **"Do spirals generally spin clockwise or anti-clockwise?"** *Explain your reasoning and provide specific evidence, with sketches if necessary, to support your reasoning.*

Group 1 results	Clockwise spiral	Anti-clockwise spiral	Edge-on/unclear
	8	4	9

16. a) Evidence-based conclusion:

Do the same as above, but for data from another group of students:

Group 2 results	Clockwise spiral	Anti-clockwise spiral	Edge-on/unclear
	19	23	8

17. b) Evidence-based conclusion:

17. c) Using complete sentences, offer at least two explanations as to why the results seem to be different between these two groups.

Phase IV: What Evidence Do You Need?

Imagine your team has been assigned the task of designing a scientific observation plan for your favorite news blog about galaxies that collide and merge into a single, larger galaxy. Describe precisely what evidence you would need to collect in order to answer the research question of, **"What fraction of galaxies observed appear to be in the process of merging with other galaxies?"**

17. *Create a detailed, step-by-step description of evidence that needs to be collected and a complete explanation of how this could be done—not just "look and see how many are merging," but exactly what would someone need to do, step-by-step, to accomplish this. You might include a table and sketches-the goal is to be precise and detailed enough that someone else could follow your procedure.*

Phase V: Formulate a Question, Pursue Evidence, and Justify Your Conclusion

Your task is to design an answerable research question, propose a plan to pursue evidence, collect data using *GalaxyZoo* (or another suitable source pre-approved by your lab instructor), and create an evidence-based conclusion about the nature and/or frequency of galaxies we observe that we have not yet addressed. Before beginning this Phase, please read through Appendix A of this lab to learn how to take quantitative measurements of the galaxies you observe. *If you want to ask a research question about the **brightness, color, or distance** of galaxies in this project, you must use the methods described in Appendix A.* There is a spare *GalaxyZoo* tally sheet on the next page if you need it to collect data.

Specific Research Question:

Step-by-Step Procedure to Collect Evidence:

Data Table and/or Results:

Evidence-based Conclusion Statement:

Additional *GalaxyZoo1* Data Table:

	Circle One		
Image	Clockwise Spiral	Anticlockwise Spiral	Edge On / Unclear
	Elliptical Galaxy		
	Star / Don't Know	Merging Galaxies	
Image	Clockwise Spiral	Anticlockwise Spiral	Edge On / Unclear
	Elliptical Galaxy		
	Star / Don't Know	Merging Galaxies	
Image	Clockwise Spiral	Anticlockwise Spiral	Edge On / Unclear
	Elliptical Galaxy		
	Star / Don't Know	Merging Galaxies	
Image	Clockwise Spiral	Anticlockwise Spiral	Edge On / Unclear
	Elliptical Galaxy		
	Star / Don't Know	Merging Galaxies	
Image	Clockwise Spiral	Anticlockwise Spiral	Edge On / Unclear
	Elliptical Galaxy		
	Star / Don't Know	Merging Galaxies	
Image	Clockwise Spiral	Anticlockwise Spiral	Edge On / Unclear
	Elliptical Galaxy		
	Star / Don't Know	Merging Galaxies	
Image	Clockwise Spiral	Anticlockwise Spiral	Edge On / Unclear
	Elliptical Galaxy		
	Star / Don't Know	Merging Galaxies	
Image	Clockwise Spiral	Anticlockwise Spiral	Edge On / Unclear
	Elliptical Galaxy		
	Star / Don't Know	Merging Galaxies	
Image	Clockwise Spiral	Anticlockwise Spiral	Edge On / Unclear
	Elliptical Galaxy		
	Star / Don't Know	Merging Galaxies	
Image	Clockwise Spiral	Anticlockwise Spiral	Edge On / Unclear
	Elliptical Galaxy		
	Star / Don't Know	Merging Galaxies	
Image	Clockwise Spiral	Anticlockwise Spiral	Edge On / Unclear
	Elliptical Galaxy		
	Star / Don't Know	Merging Galaxies	

Additional *GalaxyZoo1* Data Table:

	Circle One		
Image	Clockwise Spiral	Anticlockwise Spiral	Edge On / Unclear
	Elliptical Galaxy		
	Star / Don't Know		Merging Galaxies
Image	Clockwise Spiral	Anticlockwise Spiral	Edge On / Unclear
	Elliptical Galaxy		
	Star / Don't Know		Merging Galaxies
Image	Clockwise Spiral	Anticlockwise Spiral	Edge On / Unclear
	Elliptical Galaxy		
	Star / Don't Know		Merging Galaxies
Image	Clockwise Spiral	Anticlockwise Spiral	Edge On / Unclear
	Elliptical Galaxy		
	Star / Don't Know		Merging Galaxies
Image	Clockwise Spiral	Anticlockwise Spiral	Edge On / Unclear
	Elliptical Galaxy		
	Star / Don't Know		Merging Galaxies
Image	Clockwise Spiral	Anticlockwise Spiral	Edge On / Unclear
	Elliptical Galaxy		
	Star / Don't Know		Merging Galaxies
Image	Clockwise Spiral	Anticlockwise Spiral	Edge On / Unclear
	Elliptical Galaxy		
	Star / Don't Know		Merging Galaxies
Image	Clockwise Spiral	Anticlockwise Spiral	Edge On / Unclear
	Elliptical Galaxy		
	Star / Don't Know		Merging Galaxies
Image	Clockwise Spiral	Anticlockwise Spiral	Edge On / Unclear
	Elliptical Galaxy		
	Star / Don't Know		Merging Galaxies
Image	Clockwise Spiral	Anticlockwise Spiral	Edge On / Unclear
	Elliptical Galaxy		
	Star / Don't Know		Merging Galaxies
Image	Clockwise Spiral	Anticlockwise Spiral	Edge On / Unclear
	Elliptical Galaxy		
	Star / Don't Know		Merging Galaxies

Phase VI: Summary

18. Create a 50-word summary, in your own words, that describes the nature and frequency of galaxies we observe in the universe. You should cite specific evidence you have collected in your description, not describe what you have learned in class or elsewhere. Feel free to create and label sketches to illustrate your response.

Appendix A: How to Collect Quantitative Data for your Galaxies

To view more quantitative information on an object, click on the galaxy reference number above the buttons to classify the galaxy. This will take you to the object's reference page from the **Sloan Digital Sky Survey**, the astronomical survey that took the picture. On it is some very useful information, some of which is described below. *If you want to ask a research question about the **brightness, color, or distance** of galaxies in this project, you need to use these methods!*

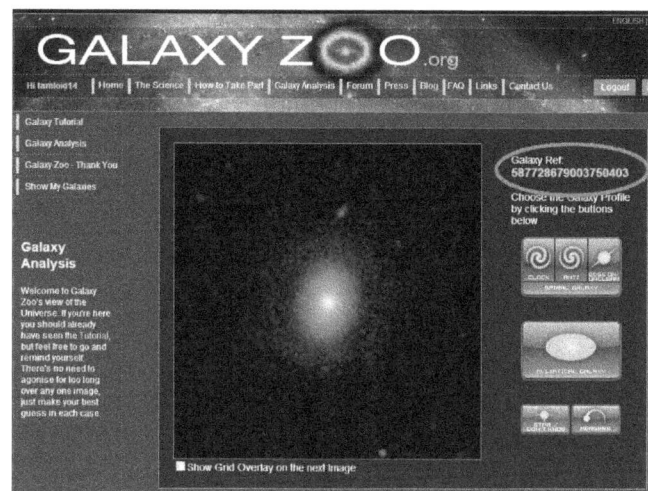

Next to the object's picture, there will be a table whose first row has column labels, "u, g, r, i, z." These are the apparent brightnesses measured in the different wavelengths bands **(bandpasses)** of the SDSS telescope. They correspond to:

u	g	r	i	z
Ultraviolet	Visible - Blue	Visible - Red	Near Infrared	Infrared

Brightness: The apparent (observed) brightness of a galaxy is measured in **magnitudes**, which is a logarithmic brightness unit. The smaller a galaxy's magnitude is in any band, the brighter the galaxy is. Conversely, the larger a galaxy's magnitude in any band, the dimmer the galaxy appears to be in that bandpass.

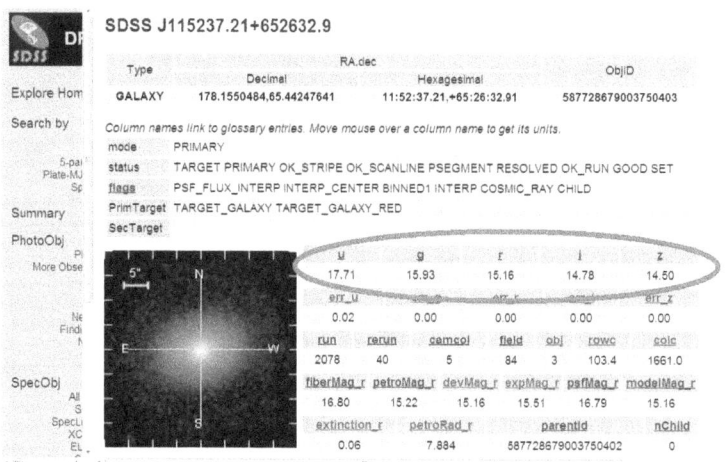

Color: The magnitudes given here can also be combined in a special way to quantitatively measure the **color** of a galaxy. Color measurements are made by subtracting the magnitude measurements in two adjacent bandpasses. So, the colors that you can calculate for these galaxies are: u - g, g - r, r - i, and i - z.

How to interpret a color: A number less than 0 for color will tell you that the galaxy has about the same brightness in both bandpasses, or is brighter in the first of your chosen bandpasses, which is more towards the *blue* side of the EM spectrum. A smaller number is very blue. A number greater than 0 for color will tell you that the galaxy is brighter in the second of the two bandpasses, which is more towards the *red* side of the EM spectrum. A larger number is very red.

A common procedure is to find the g - r color of a galaxy, which will tell you whether it appears more blue or more red in the visible part of the spectrum. Another interesting measurement is the u - g color, which will tell you whether the galaxy emits more in the ultraviolet than the visible.

Another common practice is to make a **scatter plot** with a particular color of a set of galaxies on the horizontal axis and the magnitude of the galaxies on the vertical, which creates a **color-magnitude diagram.** For instance, plotting g - r on the x-axis, and g magnitude on the y-axis. **If you do this, make sure that the magnitude you choose is one used to find the color.**

Distance: Underneath the picture of the galaxy there will most likely be another table of values with one row, and underneath that will a picture with a squiggly line. The squiggly line is the **spectrum** of the object, which has information about the composition of the galaxy and the **doppler shift** of the galaxy. Because most of these galaxies are so far away from us, the doppler shift that we measure is the **cosmological redshift** due to the expansion of the universe, which is a measure of distance of that galaxy to us. If the galaxy has an image of the spectrum listed, it will also have a measurement of the redshift. This measurement can be found in the one-row table above the image of the spectrum, in the fourth column, which is labeled "z." **DO NOT CONFUSE THIS WITH z MAGNITUDE!** A redshift near 0 is close to us, and a redshift greater than 1 is far from us. You will most likely not find any galaxy with a redshift larger than 5, so you should use that information to determine whether a galaxy is close or far from us.

SpecObjID = 168546733553352704

plate	mjd	fiberId	z	zErr	zConf	specClass	ra	dec	fiberMag_r	objId
598	52316	470	0.021	0.00016	0.999621	GALAXY	178.15510	65.44251	16.66	58772867900375040

zStatus	XCORR_EMLINE
zWarning	OK
PrimTarget	TARGET_GALAXY TARGET_GALAXY_RED
SecTarget	
eClass	-0.153157
emZ	0.018
emConf	0.054572
xcZ	0.021
xcConf	0.999621

10   **Exploring GalaxyZoo - Two**

Big Idea: The countless and varied galaxies observed throughout the Universe have characteristics that can be classified.

Goal: Students will conduct a structured series of scaffolded scientific inquiries about the nature of observed galaxies using Galaxy Zoo, a citizen science project dedicated to classifying galaxies.

Phase I: Tutorial

Access http://zoo3.galaxyzoo.org/, select HOW TO TAKE PART and complete the online GALAXY ZOO 3 TUTORIAL, completing the tables below as you go along.

First record your answers in the blocks below for each section. Then, check your answers and, in the event your answer does not agree, mark a single line through your response to help you keep track. *There is no penalty for having an incorrect answer*.

1.   Q: Is the galaxy simply smooth and rounded, with no sign of a disk? (*circle one*)

Smooth	Smooth	Smooth	Smooth	Smooth
Features or Disk	Features or Disk	Features or Disk	Features or Disk	Features or Disk
Star or Artifact	Star or Artifact	Star or Artifact	Star or Artifact	Star or Artifact
Smooth	Smooth	Smooth	Smooth	Smooth
Features or Disk	Features or Disk	Features or Disk	Features or Disk	Features or Disk
Star or Artifact	Star or Artifact	Star or Artifact	Star or Artifact	Star or Artifact

2.   Q: How rounded is it? (*circle one*)

Completely Round	Completely Round	Completely Round	Completely Round	Completely Round
In Between	In Between	In Between	In Between	In Between
Cigar Shaped	Cigar Shaped	Cigar Shaped	Cigar Shaped	Cigar Shaped
Completely Round	Completely Round	Completely Round	Completely Round	Completely Round
In Between	In Between	In Between	In Between	In Between
Cigar Shaped	Cigar Shaped	Cigar Shaped	Cigar Shaped	Cigar Shaped
Completely Round	Completely Round	Completely Round	Completely Round	Completely Round
In Between	In Between	In Between	In Between	In Between
Cigar Shaped	Cigar Shaped	Cigar Shaped	Cigar Shaped	Cigar Shaped

3. Q: Could this be a disk viewed edge-on? (*circle one*)

Yes, edge-on No, not edge-on	Yes, edge-on No, not edge-on	Yes, edge-on No, not edge-on	Yes, edge-on No, not edge-on	Yes, edge-on No, not edge-on
Yes, edge-on No, not edge-on	Yes, edge-on No, not edge-on	Yes, edge-on No, not edge-on	Yes, edge-on No, not edge-on	Yes, edge-on No, not edge-on

4. Q: Does the galaxy have a bulge at its centre? If so, what shape? (*circle one*)

Rounded Boxy No Bulge	Rounded Boxy No Bulge	Rounded Boxy No Bulge	Rounded Boxy No Bulge	Rounded Boxy No Bulge
Rounded Boxy No Bulge	Rounded Boxy No Bulge	Rounded Boxy No Bulge	Rounded Boxy No Bulge	Rounded Boxy No Bulge

5. Q: Is there any sign of a spiral arm pattern? (*circle one*)

Spiral No Spiral	Spiral No Spiral	Spiral No Spiral	Spiral No Spiral	Spiral No Spiral
Spiral No Spiral	Spiral No Spiral	Spiral No Spiral	Spiral No Spiral	Spiral No Spiral

6. Q: How tightly wound do the spiral arms appear? (*circle one*)

Tight Medium Loose	Tight Medium Loose	Tight Medium Loose	Tight Medium Loose	Tight Medium Loose
Tight Medium Loose	Tight Medium Loose	Tight Medium Loose	Tight Medium Loose	Tight Medium Loose
Tight Medium Loose	Tight Medium Loose	Tight Medium Loose	Tight Medium Loose	Tight Medium Loose

7. Q: How many spiral arms are there? (*circle one*)

1 2 3 4 >4 ?	1 2 3 4 >4 ?	1 2 3 4 >4 ?	1 2 3 4 >4 ?	1 2 3 4 >4 ?
1 2 3 4 >4 ?	1 2 3 4 >4 ?	1 2 3 4 >4 ?	1 2 3 4 >4 ?	1 2 3 4 >4 ?
1 2 3 4 >4 ?	1 2 3 4 >4 ?	1 2 3 4 >4 ?	1 2 3 4 >4 ?	1 2 3 4 >4 ?

8. Q: Is there a sign of a bar feature through the center of the galaxy? (*circle one*)

Bar No Bar	Bar No Bar	Bar No Bar	Bar No Bar	Bar No Bar
Bar No Bar	Bar No Bar	Bar No Bar	Bar No Bar	Bar No Bar

9. Q: How prominent is the central bulge, compared with the rest of the galaxy?

No Bulge Just Noticeable Obvious Dominant	No Bulge Just Noticeable Obvious Dominant	No Bulge Just Noticeable Obvious Dominant	No Bulge Just Noticeable Obvious Dominant	No Bulge Just Noticeable Obvious Dominant
No Bulge Just Noticeable Obvious Dominant	No Bulge Just Noticeable Obvious Dominant	No Bulge Just Noticeable Obvious Dominant	No Bulge Just Noticeable Obvious Dominant	No Bulge Just Noticeable Obvious Dominant
No Bulge Just Noticeable Obvious Dominant	No Bulge Just Noticeable Obvious Dominant	No Bulge Just Noticeable Obvious Dominant	No Bulge Just Noticeable Obvious Dominant	No Bulge Just Noticeable Obvious Dominant

10. Q: Can you identify an odd feature: a ring or an arc, or is the galaxy disturbed or irregular or is there a merger going on? (*circle one*)

Ring	Arc	Ring	Arc	Ring	Arc	Ring	Arc	Ring	Arc
Disturbed	Irregular	Disturbed	Irregular	Disturbed	Irregular	Disturbed	Irregular	Disturbed	Irregular
Other	Merger	Other	Merger	Other	Merger	Other	Merger	Other	Merger
Dust Lane		Dust Lane		Dust Lane		Dust Lane		Dust Lane	
Ring	Arc	Ring	Arc	Ring	Arc	Ring	Arc	Ring	Arc
Disturbed	Irregular	Disturbed	Irregular	Disturbed	Irregular	Disturbed	Irregular	Disturbed	Irregular
Other	Merger	Other	Merger	Other	Merger	Other	Merger	Other	Merger
Dust Lane		Dust Lane		Dust Lane		Dust Lane		Dust Lane	
Ring	Arc	Ring	Arc	Ring	Arc	Ring	Arc	Ring	Arc
Disturbed	Irregular	Disturbed	Irregular	Disturbed	Irregular	Disturbed	Irregular	Disturbed	Irregular
Other	Merger	Other	Merger	Other	Merger	Other	Merger	Other	Merger
Dust Lane		Dust Lane		Dust Lane		Dust Lane		Dust Lane	

[Note: more detailed descriptions of these characteristics are defined is available online at:
http://zoo3.galaxyzoo.org/how_to_take_part]

Now you are ready to move forward!

11. Go to http://www.galaxyzoo.org, select CLASSIFY to start making your own observations. Make 10 observations and circle the appropriate response. *Ignore boxes that do not apply.*

Image #	Smooth Features/Disk Star/Artifact	Round/ In between/ Cigar shape	Edge-on / Not edge-on	Round/ Boxy/ No bulge	Spiral / No spiral	Tight arms/ Medium/ Loose	Number	Bar / No Bar	No bulge / Noticeable / Obvious / Dominant	Notes

Image #	Smooth Features/Disk Star/Artifact	Round/ In between/ Cigar shape	Edge-on / Not edge-on	Round/ Boxy/ No bulge	Spiral / No spiral	Tight arms/ Medium/ Loose	Number	Bar / No Bar	No bulge / Noticeable / Obvious / Dominant	Notes

Image #	Smooth Features/Disk Star/Artifact	Round/ In between/ Cigar shape	Edge-on / Not edge-on	Round/ Boxy/ No bulge	Spiral / No spiral	Tight arms/ Medium/ Loose	Number	Bar / No Bar	No bulge / Noticeable / Obvious / Dominant	Notes

Image #	Smooth Features/Disk Star/Artifact	Round/ In between/ Cigar shape	Edge-on / Not edge-on	Round/ Boxy/ No bulge	Spiral / No spiral	Tight arms/ Medium/ Loose	Number	Bar / No Bar	No bulge / Noticeable / Obvious / Dominant	Notes

Image #	Smooth Features/Disk Star/Artifact	Round/ In between/ Cigar shape	Edge-on / Not edge-on	Round/ Boxy/ No bulge	Spiral / No spiral	Tight arms/ Medium/ Loose	Number	Bar / No Bar	No bulge / Noticeable / Obvious / Dominant	Notes

Image #	Smooth Features/Disk Star/Artifact	Round/ In between/ Cigar shape	Edge-on / Not edge-on	Round/ Boxy/ No bulge	Spiral / No spiral	Tight arms/ Medium/ Loose	Number	Bar / No Bar	No bulge / Noticeable / Obvious / Dominant	Notes

Image #	Smooth Features/Disk Star/Artifact	Round/ In between/ Cigar shape	Edge-on / Not edge-on	Round/ Boxy/ No bulge	Spiral / No spiral	Tight arms/ Medium/ Loose	Number	Bar / No Bar	No bulge / Noticeable / Obvious / Dominant	Notes

Image #	Smooth Features/Disk Star/Artifact	Round/ In between/ Cigar shape	Edge-on / Not edge-on	Round/ Boxy/ No bulge	Spiral / No spiral	Tight arms/ Medium/ Loose	Number	Bar / No Bar	No bulge / Noticeable / Obvious / Dominant	Notes

Image #	Smooth Features/Disk Star/Artifact	Round/ In between/ Cigar shape	Edge-on / Not edge-on	Round/ Boxy/ No bulge	Spiral / No spiral	Tight arms/ Medium/ Loose	Number	Bar / No Bar	No bulge / Noticeable / Obvious / Dominant	Notes

Image #	Smooth Features/Disk Star/Artifact	Round/ In between/ Cigar shape	Edge-on / Not edge-on	Round/ Boxy/ No bulge	Spiral / No spiral	Tight arms/ Medium/ Loose	Number	Bar / No Bar	No bulge / Noticeable / Obvious / Dominant	Notes

12. Rate the relative DIFFICULTY your team has distinguishing the following by circling one.

Rate the difficulty of classifying each of the following:	Nearly Impossible	Challenging	Some not easy, Varies	Pretty Easy	Notes or Comments
Presence of Spiral Arms	Imposs.	Chall.	Varies	Easy	
Roundness of Galaxies	Imposs.	Chall.	Varies	Easy	
Tightness of Spiral Arms	Imposs.	Chall.	Varies	Easy	
Number of Spiral Arms	Imposs.	Chall.	Varies	Easy	
Evidence of Central Bar	Imposs.	Chall.	Varies	Easy	
Dominance of Central Bulge	Imposs.	Chall.	Varies	Easy	

Phase II: Does the Evidence Match the Conclusion?

Enter the CLASSIFY Galaxy Zoo scientific database and classify ten (10) additional images. Keep a record of your results in the "Tally Sheet" below using tick marks. ⅢⅡ

Note: Not all options will appear for every galaxy. The options that appear are dependent on the ones that you've already selected. Be sure to mark only the options that you click for each galaxy.

TALLY SHEET Data Table "A"	Features/ Disk	Round	Edge-on	Round	Spiral	Tight arms	Bar	No bulge
		In between		Boxy		Medium		Noticeable
	Smooth	Cigar shape	Not edge-on	No bulge	No spiral	Loose	No Bar	Obvious
								Dominant

13. Consider the research question, "which shape of elliptical galaxy is most common?" If a student proposed a generalization that "**most elliptical galaxies are cigar-shaped**," would you agree or disagree with the generalization based on all the evidence you collected so far? Pursue this evidence by considering how many galaxies are cigar-shaped compared to the total number of elliptical galaxies you have observed. *Explain your reasoning and provide specific evidence either from the above questions or from additional evidence you yourself generate using GalaxyZoo.*

Phase III: What Conclusions Can You Draw From the Evidence?

Galaxies are observed to have numerous different shapes. What conclusions and generalizations can you make from the following data collected by a student in terms of the question, **"Do spiral galaxies generally exhibit a clear central bulge?"** *Explain your reasoning and provide specific evidence, with sketches if necessary, to support your reasoning.*

Bulge Data Table	No Bulge	Noticeable Bulge	Obvious Bulge	Dominant Bulge
Sample #1	4	17	19	6
Sample #2	10	21	14	1
Sample #3	7	23	12	6

14. Evidence-based conclusion:

Phase IV: What Evidence Do You Need?

15. Imagine your team has been assigned the task of designing a scientific observation plan for your favorite news blog about number of spiral arms a galaxy has. Describe precisely what evidence you would need to collect in order to answer the research question of, "**How many arms do spiral galaxies have?**"

    Create a detailed, step-by-step description of evidence that needs to be collected and a complete explanation of how this could be done—not just "look and see how many arms are there," but exactly what would someone need to do, step-by-step, to accomplish this. You might include a table and sketches-the goal is to be precise and detailed enough that someone else could follow your procedure. Do NOT include generic nonspecific steps such as "analyze data" or "present conclusions" -- these are meaningless filler. Be specific!

Phase V: Formulate a Question, Pursue Evidence, and Justify Your Conclusion

Your task is to design an answerable research question, propose a plan to pursue evidence, collect data using *GalaxyZoo* (or another suitable source pre-approved by your lab instructor), and create an evidence-based conclusion about the frequency of observable characteristics of galaxies that we have not yet addressed.

Specific Research Question:

Step-by-Step Procedure to Collect Evidence:

Data Table and/or Results:

Evidence-based Conclusion Statement:

Phase VI: Summary

Create a 50-word summary, in your own words, that describes the nature and characteristics of galaxies we observe in the universe. You should cite specific evidence you have collected in your description, not describe what you have learned in class or elsewhere. Feel free to reference specific parts of this lab or create and label sketches to illustrate your response.

Additional GalaxyZoo 3 Data Sheet

Image #	Smooth Features/Disk Star/Artifact	Round/ In between/ Cigar shape	Edge-on / Not edge-on	Round/ Boxy/ No bulge	Spiral / No spiral	Tight arms/ Medium/ Loose	Number	Bar / No Bar	No bulge / Noticeable / Obvious / Dominant	Notes
Image #	Smooth Features/Disk Star/Artifact	Round/ In between/ Cigar shape	Edge-on / Not edge-on	Round/ Boxy/ No bulge	Spiral / No spiral	Tight arms/ Medium/ Loose	Number	Bar / No Bar	No bulge / Noticeable / Obvious / Dominant	Notes
Image #	Smooth Features/Disk Star/Artifact	Round/ In between/ Cigar shape	Edge-on / Not edge-on	Round/ Boxy/ No bulge	Spiral / No spiral	Tight arms/ Medium/ Loose	Number	Bar / No Bar	No bulge / Noticeable / Obvious / Dominant	Notes
Image #	Smooth Features/Disk Star/Artifact	Round/ In between/ Cigar shape	Edge-on / Not edge-on	Round/ Boxy/ No bulge	Spiral / No spiral	Tight arms/ Medium/ Loose	Number	Bar / No Bar	No bulge / Noticeable / Obvious / Dominant	Notes
Image #	Smooth Features/Disk Star/Artifact	Round/ In between/ Cigar shape	Edge-on / Not edge-on	Round/ Boxy/ No bulge	Spiral / No spiral	Tight arms/ Medium/ Loose	Number	Bar / No Bar	No bulge / Noticeable / Obvious / Dominant	Notes
Image #	Smooth Features/Disk Star/Artifact	Round/ In between/ Cigar shape	Edge-on / Not edge-on	Round/ Boxy/ No bulge	Spiral / No spiral	Tight arms/ Medium/ Loose	Number	Bar / No Bar	No bulge / Noticeable / Obvious / Dominant	Notes
Image #	Smooth Features/Disk Star/Artifact	Round/ In between/ Cigar shape	Edge-on / Not edge-on	Round/ Boxy/ No bulge	Spiral / No spiral	Tight arms/ Medium/ Loose	Number	Bar / No Bar	No bulge / Noticeable / Obvious / Dominant	Notes
Image #	Smooth Features/Disk Star/Artifact	Round/ In between/ Cigar shape	Edge-on / Not edge-on	Round/ Boxy/ No bulge	Spiral / No spiral	Tight arms/ Medium/ Loose	Number	Bar / No Bar	No bulge / Noticeable / Obvious / Dominant	Notes
Image #	Smooth Features/Disk Star/Artifact	Round/ In between/ Cigar shape	Edge-on / Not edge-on	Round/ Boxy/ No bulge	Spiral / No spiral	Tight arms/ Medium/ Loose	Number	Bar / No Bar	No bulge / Noticeable / Obvious / Dominant	Notes
Image #	Smooth Features/Disk Star/Artifact	Round/ In between/ Cigar shape	Edge-on / Not edge-on	Round/ Boxy/ No bulge	Spiral / No spiral	Tight arms/ Medium/ Loose	Number	Bar / No Bar	No bulge / Noticeable / Obvious / Dominant	Notes
Image #	Smooth Features/Disk Star/Artifact	Round/ In between/ Cigar shape	Edge-on / Not edge-on	Round/ Boxy/ No bulge	Spiral / No spiral	Tight arms/ Medium/ Loose	Number	Bar / No Bar	No bulge / Noticeable / Obvious / Dominant	Notes
Image #	Smooth Features/Disk Star/Artifact	Round/ In between/ Cigar shape	Edge-on / Not edge-on	Round/ Boxy/ No bulge	Spiral / No spiral	Tight arms/ Medium/ Loose	Number	Bar / No Bar	No bulge / Noticeable / Obvious / Dominant	Notes

Image #	Smooth Features/Disk Star/Artifact	Round/ In between/ Cigar shape	Edge-on / Not edge-on	Round/ Boxy/ No bulge	Spiral / No spiral	Tight arms/ Medium/ Loose	Number	Bar / No Bar	No bulge / Noticeable / Obvious / Dominant	Notes
Image #	Smooth Features/Disk Star/Artifact	Round/ In between/ Cigar shape	Edge-on / Not edge-on	Round/ Boxy/ No bulge	Spiral / No spiral	Tight arms/ Medium/ Loose	Number	Bar / No Bar	No bulge / Noticeable / Obvious / Dominant	Notes

Image #	Smooth Features/Disk Star/Artifact	Round/ In between/ Cigar shape	Edge-on / Not edge-on	Round/ Boxy/ No bulge	Spiral / No spiral	Tight arms/ Medium/ Loose	Number	Bar / No Bar	No bulge / Noticeable / Obvious / Dominant	Notes

Image #	Smooth Features/Disk Star/Artifact	Round/ In between/ Cigar shape	Edge-on / Not edge-on	Round/ Boxy/ No bulge	Spiral / No spiral	Tight arms/ Medium/ Loose	Number	Bar / No Bar	No bulge / Noticeable / Obvious / Dominant	Notes

Image #	Smooth Features/Disk Star/Artifact	Round/ In between/ Cigar shape	Edge-on / Not edge-on	Round/ Boxy/ No bulge	Spiral / No spiral	Tight arms/ Medium/ Loose	Number	Bar / No Bar	No bulge / Noticeable / Obvious / Dominant	Notes

Image #	Smooth Features/Disk Star/Artifact	Round/ In between/ Cigar shape	Edge-on / Not edge-on	Round/ Boxy/ No bulge	Spiral / No spiral	Tight arms/ Medium/ Loose	Number	Bar / No Bar	No bulge / Noticeable / Obvious / Dominant	Notes

Image #	Smooth Features/Disk Star/Artifact	Round/ In between/ Cigar shape	Edge-on / Not edge-on	Round/ Boxy/ No bulge	Spiral / No spiral	Tight arms/ Medium/ Loose	Number	Bar / No Bar	No bulge / Noticeable / Obvious / Dominant	Notes

Image #	Smooth Features/Disk Star/Artifact	Round/ In between/ Cigar shape	Edge-on / Not edge-on	Round/ Boxy/ No bulge	Spiral / No spiral	Tight arms/ Medium/ Loose	Number	Bar / No Bar	No bulge / Noticeable / Obvious / Dominant	Notes

Image #	Smooth Features/Disk Star/Artifact	Round/ In between/ Cigar shape	Edge-on / Not edge-on	Round/ Boxy/ No bulge	Spiral / No spiral	Tight arms/ Medium/ Loose	Number	Bar / No Bar	No bulge / Noticeable / Obvious / Dominant	Notes

Image #	Smooth Features/Disk Star/Artifact	Round/ In between/ Cigar shape	Edge-on / Not edge-on	Round/ Boxy/ No bulge	Spiral / No spiral	Tight arms/ Medium/ Loose	Number	Bar / No Bar	No bulge / Noticeable / Obvious / Dominant	Notes

Image #	Smooth Features/Disk Star/Artifact	Round/ In between/ Cigar shape	Edge-on / Not edge-on	Round/ Boxy/ No bulge	Spiral / No spiral	Tight arms/ Medium/ Loose	Number	Bar / No Bar	No bulge / Noticeable / Obvious / Dominant	Notes

Big Idea: Designing a fruitful plan for conducting research has many pitfalls. By assessing the research reports of others, scientists can improve their own ability to design attractive research plans. With better research designs, researchers can improve the support for the claims they make with better and better evidence.

Goal: Students will assess a series of research reports and then select one project to redesign and conduct in order to more productively pursue the original research question.

Assess Research Projects and Identify Inconsistencies in Their Lines of Inquiry:

In this lab you will identify any issues with research projects similar to those you have already completed. Your task is to distinguish between severe problems that could invalidate the research, and those that are minor. Work on improving only one research report at a time. Make sure to specify which report you are using by completely writing out the research question. Answer each of the questions by providing a short, but detailed, explanation of your reasoning citing specific information from the provided research reports.

Inquiry Research Report #21
Investigating Exoplanets

Formulate a Question, Pursue Evidence, and Justify Your Conclusion

Your task is to design an answerable research question, propose a plan to pursue evidence, collect data using exoplanets.org (or another suitable source pre-approved by your lab instructor), and create an evidence-based conclusion about properties of exoplanets that you have not completed before.

Research report:

Specific research question:

Is the mass of an exoplanet correlated with the mass of the star that it orbits?

Step-by-step procedure to collect evidence:

Using the Exoplanets Data Explorer – Plots at exoplanets.org:
1. Select Scatter Plot of the following two properties: masses of all known exoplanets, and the masses of the stars that these planets orbit.
2. Plot the masses of each of the exoplanets against the mass of the stars they orbit. This means the y-axis is exoplanet mass in terms of Jupiter's mass, and x-axis is the mass of the stars that exoplanets orbit in terms of the Sun's mass.

Data table and/or results:

See graph at right:

Evidence-based conclusion statement:

Only stars that have a mass very similar to the mass of the Sun have planets orbiting them.

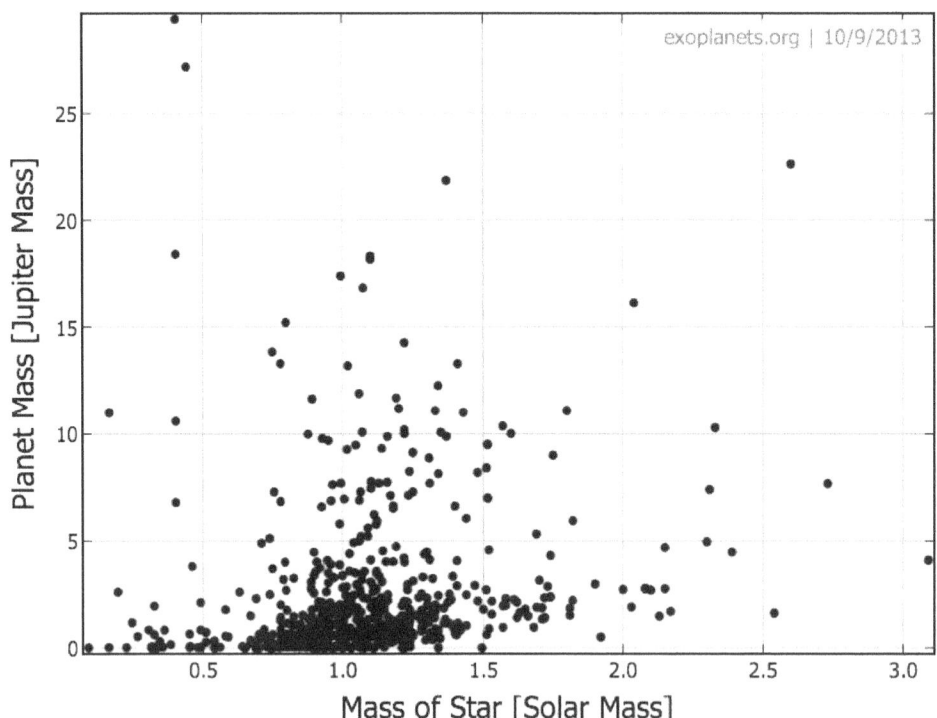

Report analysis

Report number: _____

Keywords from Research question: _____

1) Write down some of the things you might observe to pursue this research question. *If you already took notes while reading the report, then there is no need to copy your notes here.*

2) List any problems with this report. **Make separate lists** for *minor* problems and *major* problems. *You may want to consider the following questions in determining whether something is a major or a minor problem. (You can use the space in the previous question to make notes.)*

   a) *Did they collect relevant evidence?*
   b) *Have they collected enough evidence? Or is their evidence insufficient and anecdotal?*
   c) *Did they claim more than the evidence supports?*
   d) *Did they follow their procedure?*
   e) *Do they answer the research question?*
   f) *Have assumptions impacted their results? That is, have the researchers made use of unjustified prior knowledge in lieu of collecting data?*

3) Do the major problems invalidate the research? *If no, explain why you classified them as major rather than minor.*

4) Do the minor problems invalidate the research? *If yes, explain why you classified them this way.*

5) Is the presentation of their results clear and unambiguous? What about the rest of the report? Anything that stands out as good or bad in their presentation of the report? *(If you already identified these problems as minor or major, just reference them here. No need to repeat yourself.)*

6) Precisely what should the researchers have done or reported differently to improve their research project?

Inquiry Research Report #22
Observing the Moon

Formulate a Question, Pursue Evidence, and Justify Your Conclusion

Your task is to design an answerable research question, propose a plan to pursue evidence, collect data using Solar System Simulator (or another suitable source pre-approved by your lab instructor), and create an evidence-based conclusion about the orbit or motion of a planet or moon that you have not completed before.

Research report:

Specific research question:

Exactly how many days is the Moon's orbital period around the Earth?

Step-by-step procedure to collect evidence:

Using the Solar System Simulator at http://space.jpl.nasa.gov (observe from Sun's vantage point):
1. Beginning Feb. 14, 2009, observe the Moon/Earth system every 4 days for 3 months.
2. For each observation, measure the distance between the centers of Earth and Moon.
3. Making measurements with a ruler (in any units – Just be consistent) mark down the distance and on which side of the Earth the Moon is located, left or right.
4. Record measurements in a table, and plot the data in a graph of Distance vs. Time to determine the orbital period of the Moon around Earth.

Data table and/or results:

See next page

Evidence-based conclusion statement:

Exactly 30 days

Date	Distance	Left/Right
14-Feb	10	L
19-Feb	11.75	L
24-Feb	2.75	L
1-Mar	9	R
6-Mar	10.75	R
11-Mar	1	R
16-Mar	10.75	L
21-Mar	11.5	L
26-Mar	2	L
31-Mar	9.75	R
5-Apr	10	R
10-Apr	1.5	L
15-Apr	11.25	L
20-Apr	11.25	L
25-Apr	1	L
30-Apr	10.5	R
5-May	9.25	R
10-May	2.25	L

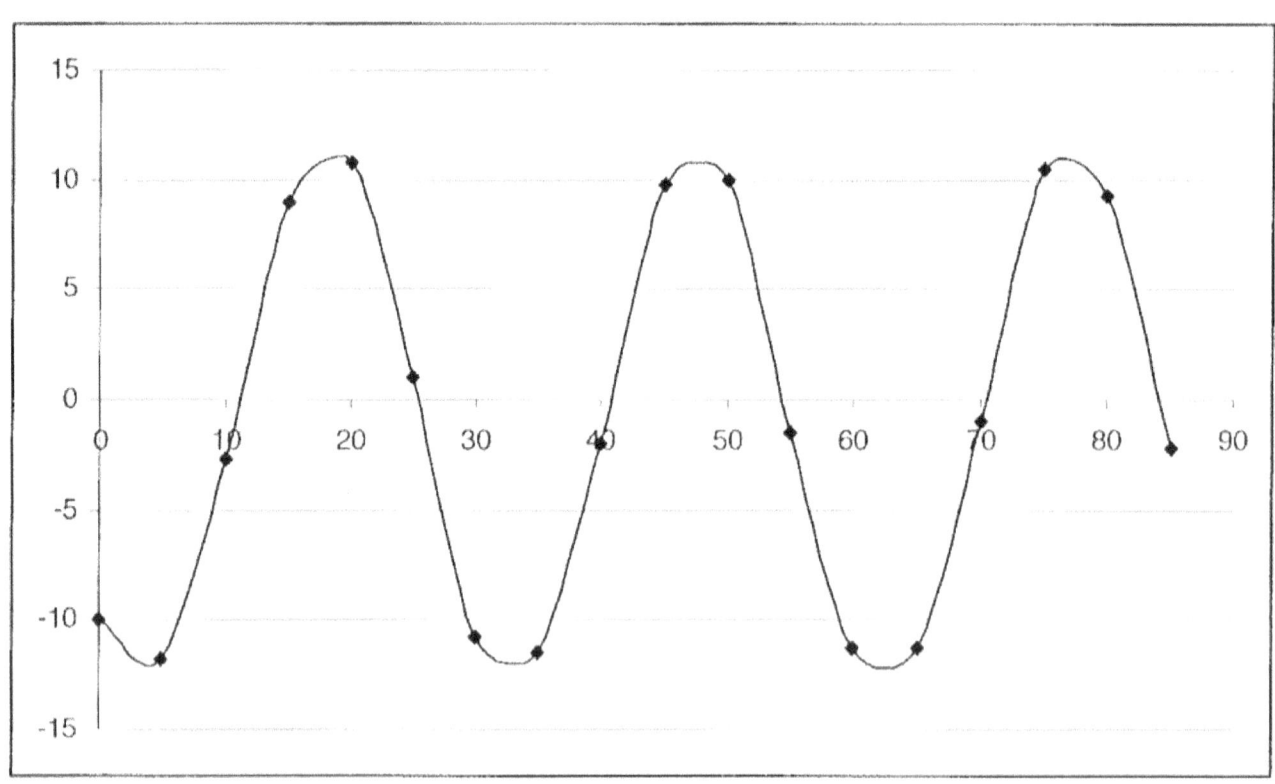

Report analysis

Report number: _____

Keywords from Research question: _____

7) Write down some of the things you might observe to pursue this research question. *If you already took notes while reading the report, then there is no need to copy your notes here.*

8) List any problems with this report. **Make separate lists** for *minor* problems and *major* problems. *You may want to consider the following questions in determining whether something is a major or a minor problem. (You can use the space in the previous question to make notes.)*

   a) *Did they collect relevant evidence?*
   b) *Have they collected enough evidence? Or is their evidence insufficient and anecdotal?*
   c) *Did they claim more than the evidence supports?*
   d) *Did they follow their procedure?*
   e) *Do they answer the research question?*
   f) *Have assumptions impacted their results? That is, have the researchers made use of unjustified prior knowledge in lieu of collecting data?*

9) Do the major problems invalidate the research? *If no, explain why you classified them as major rather than minor.*

10) Do the minor problems invalidate the research? *If yes, explain why you classified them this way.*

11) Is the presentation of their results clear and unambiguous? What about the rest of the report? Anything that stands out as good or bad in their presentation of the report? *(If you already identified these problems as minor or major, just reference them here. No need to repeat yourself.)*

12) Precisely what should the researchers have done or reported differently to improve their research project?

Inquiry Research Report #23
Classifying Exoplanets

Formulate a Question, Pursue Evidence, and Justify Your Conclusion

Your task is to design an answerable research question, propose a plan to pursue evidence, collect data using exoplanets.org, and create an evidence-based conclusion about exoplanets that you have not completed before.

Research report:

Specific research question:

Do exoplanets with orbital distances similar to that of the Earth-Sun distance also have masses similar to that of Earth's mass? (In other words, does Earth-distanced correlate to Earth-massed?)

Step-by-step procedure to collect evidence:

Using the Exoplanets Data Explorer – Plots at exoplanets.org:
1. Select Scatter Plot of the following two properties: orbital distances and planet mass.
2. Plot the masses of each of the exoplanets against their orbital distance (semi-major axis). This means the y-axis is exoplanet mass in terms of Jupiter's mass, and the x-axis is the semi-major axis in units of AU.

Data table and/or results:

See graph at right:

Evidence-based conclusion statement:

There appears to be no correlation between mass and orbital distance for planets with orbital distances similar to that of the Earth.

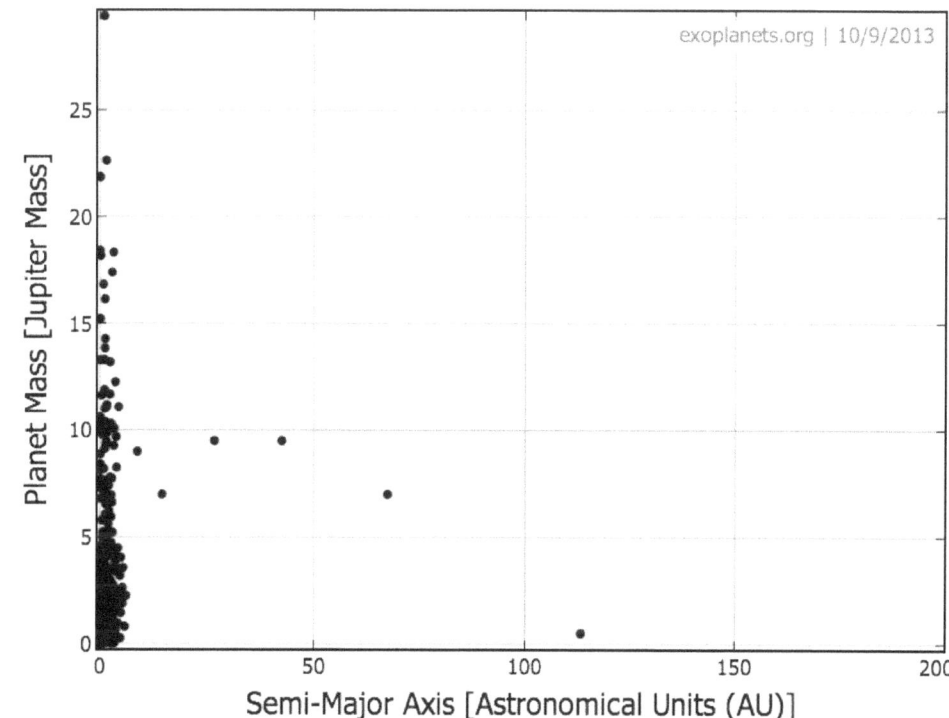

Report analysis

Report number: _____

Keywords from Research question: _____

13) Write down some of the things you might observe to pursue this research question. *If you already took notes while reading the report, then there is no need to copy your notes here.*

14) List any problems with this report. **Make separate lists** for *minor* problems and *major* problems. *You may want to consider the following questions in determining whether something is a major or a minor problem. (You can use the space in the previous question to make notes.)*

   a) *Did they collect relevant evidence?*
   b) *Have they collected enough evidence? Or is their evidence insufficient and anecdotal?*
   c) *Did they claim more than the evidence supports?*
   d) *Did they follow their procedure?*
   e) *Do they answer the research question?*
   f) *Have assumptions impacted their results? That is, have the researchers made use of unjustified prior knowledge in lieu of collecting data?*

15) Do the major problems invalidate the research? *If no, explain why you classified them as major rather than minor.*

16) Do the minor problems invalidate the research? *If yes, explain why you classified them this way.*

17) Is the presentation of their results clear and unambiguous? What about the rest of the report? Anything that stands out as good or bad in their presentation of the report? *(If you already identified these problems as minor or major, just reference them here. No need to repeat yourself.)*

18) Precisely what should the researchers have done or reported differently to improve their research project?

Choose One Research Project to Redesign, Improve, and Conduct

Your task is to choose one of the research projects (either report 22 or report 23) to redesign and carry out. You should re-use the <u>exact same research question as the previous researchers</u>, but make sure to improve the research design so that you eliminate the main problem(s) you were able to identify. Note that the next page contains an empty grid that may help you make any graphs more precisely.

Your *redesigned* research report:

Specific research question:

Step-by-step procedure to collect evidence:

Data table and/or results:

Evidence-based conclusion statement:

19) Precisely what has been done or reported differently to improve the original research inquiry project?

Summary

20) Science, what is it all about? *Write a 50-word (short and concise!) essay on what science is all about. You may want to consider the role of evidence, and questions like why it has produced such reliable knowledge, and whether the principles and methods could also be used in other areas.*

21) Why is mathematics useful?

22) Is statistics useful?

HZ	Habitable Zone

Introduction and Background: Earth is the only planet known to harbor life of any kind, past or present. As part of the search for evidence of life on other planets, both in our Solar System and in other planetary systems, we are looking not just for evidence of living organism themselves, but for evidence of the conditions that might even be hospitable to life as we know it. One feature that astronomers consider to determine whether a planet MIGHT have the conditions necessary for life as we know it is whether the planet falls within a '*habitable zone*' of its host star. (Note: This lab will address circumstellar habitable zones only, *not* galactic habitable zones.) If you have already learned about habitable zones in another astronomy class or from your own general knowledge, great. If not, or if you'd like to get a stronger background before proceeding, please read more about habitable zones at one or more of the following websites:

astro.unl.edu: http://astro.unl.edu/naap/habitablezones/chz.html
universetoday.com: http://www.universetoday.com/32622/habitable-zone/
astronomynotes.com: http://www.astronomynotes.com/lifezone/s2.htm
space.com: http://www.space.com/2021-growing-habitable-zone-locations-life-abound.html

Goal: Students will conduct a series of inquiries about the nature of circumstellar habitable zones and the factors they depend on, and the timescale for evolution of life on Earth.

Computer Setup: Access the University of Nebraska's Habitable Zone Simulator at the following URL: http://astro.unl.edu/naap/habitablezones/animations/stellarHabitableZone.swf

Phase I: The Habitability of the Earth

A habitable zone is the region around a star where the amount of light received from the star by objects in that region (namely planets) leads to temperatures in which any water on the surface could exist in liquid form. The amount of light received by a planet from a star depends on its distance from the star (farther away means less light), and therefore so does the planet's surface temperature.

1. In this definition of habitable zone...

    a) ...what is meant by "habitable"?

    b) ...why is it a "zone" and not one specific location?

    c) ...what type of astronomical object is it surrounding?

    d) ...what object(s) may be located within it?

    e) ...what is the possible temperature range for a planet in the habitable zone of its star (quantitatively)?

Load the Habitable Zone Simulator. The flash simulator will show you a visual diagram of the solar system in the top panel, a set of simulation settings in the middle panel, and a timeline of the habitability of the Earth in the bottom panel. The timeline units will either be Megayears (Myr), which means millions of years, or Gigayears (Gyr), which means billions of years. To run the simulation, click **run** in the bottom panel. This button immediately becomes a *pause* button, which will allow you to pause the simulation at any time.

The simulation runs pretty quickly by default. To adjust the speed, use the *rate* slider bar to the right of the *run* button. You can also manually advance the simulation forward or backward by clicking and dragging the upside-down dark grey triangle above the timeline. To restore the simulation to the original default settings, press the *reset* button at the very top of the simulation.

2. The simulation is currently set to zero-age - this is the Solar System as it was when it first formed, about 4.5 billion years ago. Which planet(s) were in the Habitable Zone at this time, if any? _____

3. The blue region marked on the diagram is the Habitable Zone around our Sun. Notice how there is both an inner edge and an outer edge - the planets interior to the habitable zone are too hot to support liquid water, while the planets exterior to it are too cold. Why?

4. Below are two stars and their habitable zones, with distances drawn to the same scale (stars not drawn to scale). Which star is brighter/hotter? How do you know?

Star A   Star B

5. Press the **start** button and watch the Habitable Zone change with time. Pause the simulation when it reaches an age of 4.5 billion years (age=time since star system formation; you can keep track of the time by looking at the timeline marker in the bottom panel). This is the Solar System as it is today - which planet(s) are in the Habitable Zone now, if any?

_____

6. Allow the simulation to run until the Earth is no longer in the Habitable Zone.

    a) At what age does this happen? _____

    b) How long <u>from now</u> until this happens? _____

    You can use the timeline bar in the bottom panel to determine your answers. Be sure to include units with your numbers above.

7. After the Earth is no longer within the Habitable Zone, what do you think the conditions on Earth will be like, and why?

8. Resume the simulation and let it run until the end. Which planets other than the Earth will fall within the Habitable Zone at *any point* during the Sun's life, if any?

    _____

9. Why does the habitable zone change during the Sun's lifetime? Pay attention to how the properties of the Sun change, and explain how this can affect the habitability of planets. <u>One or two full sentences please.</u>

10. *Optional Challenge Question: Around 12 billion years, the Earth's distance from the Sun suddenly changes. Why? (Draw from your knowledge of what you learned in your previous astronomy class that was a prerequisite or corequisite for taking this lab.) Extra credit is possible for very good answers.*

Phase II: The History of Life on Earth

As you saw in the simulations above, the Earth has been within the Habitable Zone of our Sun nearly since its formation 4.5 billion years ago. Complex life, however, did not develop immediately. And humans did not appear until later still. The timeline at right delineates several milestones in the history of life on Earth.

Billions of Years Ago	Significant Events
4.5	The Earth Forms
3.8	First Life
2.8	Photosynthesis
1.5	Organisms with More than One Cell
0.45	Land Animals
0.0005	Human Life

11. For each of the events on the timeline, determine how long after the formation of the Earth this event occurred (in Gigayears -- "Giga" means billion) Then, calculate what fraction of its current age (4.5 billion years) the Earth was at that time. Fill in your answers on the table below.

For example, in the 2nd row: the first primitive life arose 3.8 Gyr ago, which was 0.7 Gyr after Earth formed (4.5 Gyr - 3.8 Gyr = 0.7 Gyr). At that time, when Earth was 0.7 Gyr old, that was 0.7/4.5 = 0.155 = 15.5% of Earth's now current age.

Significant Event	Age of Earth at that time	% of Earth's current age
Earth forms	Gyr	%
First life emerges	Gyr	%
First photosynthesis	Gyr	%
Multicellular organisms	Gyr	%
Land animals	Gyr	%
First humans	Gyr	%

12. Think about your answers to the previous timeline question. What do you think was the purpose of that exercise? What is the take-home message? (Think about whether primitive life arose early or late. What about humans?) <u>One or two full sentences please.</u>

Phase III: The Habitable Zones of Different Kinds of Stars

Now that you've simulated the Habitable Zone (HZ) around our Sun, we'll run the same simulation for other stars. For seven different star types, your job will be to find the planet orbit that remains in the HZ the longest. This will take some time! This is the main part of this lab.

Astronomers classify stars with letters: O, B, A, F, G, K, and M. The O stars are the hottest and most luminous, while M stars are the coolest and dimmest. Every types of star has its own HZ, with brighter stars having more-distant HZs. Imagine putting an extra log on a campfire; the campers all have to back off a few feet to maintain the same comfortable temperature.

Below is a table of the different types of stars in the classification scheme above. Notice how they each have a different mass - in fact, the **mass** of a star is the underlying determining factor for all other stellar properties (luminosity, temperature, etc.), and therefore dictates how it will be classified. The highest-mass stars are hottest.

Reset the HZ simulator with the *reset* button at top, and then adjust the star mass with the *initial star mass* slider bar in the middle panel. The units of star mass are <u>Solar Masses</u> ($M_\odot$); our Sun's mass is exactly one Solar Mass (1.0 $M_\odot$) by definition. Notice how the HZ immediately changes in size. Notice also that you can adjust the orbit of "Earth" (i.e., the planet under consideration) by adjusting the *initial planet distance* slider bar in the middle panel. You can also adjust it by clicking on the planet itself and dragging it closer or farther from the star. The units of distance from the star are AU - astronomical units, which is defined as the real-life distance of the Earth from the Sun. The Earth is 1 AU from the Sun by definition.

13. For each type of star in the table below, run the simulator with the closest mass you can find to that listed as "typical" for that type. Indicate what mass you chose in the third column, even if it was identical to the typical mass listed in the second column. Adjust the initial planet distance (we suggest dragging the planet back and forth slowly through the HZ while keeping an eye on the total length of the blue bar, indicating time of habitability, on the bottom) until you find the one that gives the longest amount of time *continuously* in the HZ; record both the initial planet distance used and the corresponding *total* time in the HZ. Note: you are recording the *total* time continuously in the HZ for the longest stretch, *not* necessarily just the time when the planet leaves the HZ, which may be different. Finally, in the last column, record the most advanced life (if any) that could develop in this amount of time, using your answers from the table in the previous part.

For some of the lower-mass stars, you should find that the planet becomes tidally locked even while it is still in the habitable zone. **Ignore tidal locking**, *and just pay attention to when the planet is in the HZ.*

Star Type	Typical Star Mass (M$_{Sun}$)	Star Mass used in simulator (M$_{Sun}$)	Orbit Size of longest habitable orbit (AU)	Habitable Lifetime (Gyr*)	Most advanced life that could develop
O	16				
B	5				
A	2				
F	1.3				
G	1.0				
K	0.7				
M	0.4				

*WARNING: Sometimes the numbers on the timeline are shown in Myr (Megayears, where "Mega" = million) instead of Gyr (Gigayears, where "Giga" = billion) in cases where the star lives are short enough to warrant these units. Be sure to **convert times in Myr to Gyr** as necessary before you enter your answer! If you need help with this conversion, ask other students or else the professor!

14. Given your answers in the table above, and keeping in mind that the Universe is only 13.7 billion years old, what type of star do you think would be the best place to look for planets harboring life, and why? One or two full sentences, please.

15. What do you notice about the TOTAL lifetimes of the different types of stars? (That is, the lifetimes of the stars themselves, ignoring any planets and the HZ.) Which live the longest, and which the shortest?

16. a) Which type of star is most luminous? _____
    b) Which is least? _____
    c) So which is easiest to detect and monitor?

17. What type of star is our Sun? _____

18. a) Compared to our Sun's type (see above), what do you think the development of life on planets orbiting *hotter* types of stars would be like?

    b) What about *cooler* types of stars?

    Note: Do you think that life in such conditions is even possible? Either way, justify your above answers in one or two sentences each.

19. If you were the director of a NASA program to search for life beyond Earth, toward which type of star would you direct your attention, and why? Justify your answer, including evidence from previous questions. You may also use any additional lines of reasoning you like. Several full sentences please.

Phase IV: The Habitability of Different Kinds of Stars

20. Jupiter's moon Europa is currently covered with water ice (H_2O), and possibly liquid water beneath. How is this possible, given that Jupiter is well outside our Sun's current HZ? Be sure to consider on which side of our Sun's HZ Jupiter and Europa are located, and include in your answer what assumptions go into the standard definition of "habitable zone" as used by this simulator. Two or three full sentences would be appropriate.

21. If a planet or moon IS inside the habitable zone, does that necessarily mean it is habitable? Why or why not? (Hint: Earth's Moon is inside our Sun's habitable zone. Is it habitable?)

Phase V: Reflection and Conclusions

Most of the stars we can see with the unaided eye in our night sky are hundreds or even thousands of lightyears away from Earth. (The very closest ones are only a few dozen lightyears away, but most are much further.) The vast majority of stars in our galaxy are many tens of thousands of lightyears away. **IF** intelligent life existed on planets orbiting some of these stars -- and that's a huge IF! -- comment on the likelihood and practicality of (a) visiting, (b) communicating with, or (c) verifying the existence of those life forms. Describe how you might go about approaching EACH of these three tasks, or if you think they are even possible. <u>(One or two sentences for each part would be appropriate.)</u>

Your Name: _____

Group Member(s): _____

Big Idea:

Stars have a number of properties that, at first glance, may appear to be unrelated. But further analysis shows that they are related after all. Two astronomers, independently of each other, plotted the luminosities of a number of stars versus their temperatures to create what is now known as the "Hertzsprung-Russell", or "HR" diagram (named after the scientists who invented it), which allows us to study this relationship between properties.

Goal:

In this lab, you will find that for the majority of stars there is a definite relationship between temperature and luminosity. You will learn that other properties of stars are also related. In fact, the HR diagram contains a surprisingly large amount of information in one simple graph. Another goal of this lab is to review graphing skills, which are a central learning objective of the course. By the end of this activity, you will be expected to be able to read a graph and create one of your own, including clear labels and units. You should also be able to describe in words what is demonstrated by a graph of data -- not just "the points form a line" but in full sentences and including the physical meaning of the relationship.

Materials Needed:

Multicolored crayons, and a ruler or straightedge

In the two tables and graph that follow, **temperatures** are in Kelvin (K), and **luminosities** and **masses** are given in units relative to the Sun; the units are called: "Solar Luminosities" and "Solar Masses" respectively, which are abbreviated as $L_\odot$ and $M_\odot$. Note that the values for the Sun in these units are 1.0 $L_\odot$ and 1.0 $M_\odot$ by definition. **Spectral type** is a classification scheme for stars that depends on their temperature. From hottest to coldest, the sequence of spectral types goes OBAFGKM (which corresponds with the first letter of each entry in this column).

First, here is a list of the 25 **nearest** stars in the sky, starting with the nearest:

	Star	Spectral Type	Temp.	Luminosity	Mass
1	Sun	G2V	5,800	1.0	1.0
2	Proxima Cen	M5V	3,100	0.000 058	
3	α Cen A	G2V	5,800	1.60	1.10
4	α Cen B	K0V	5,000	0.46	0.85
5	Barnard's star	M3V	3,100	0.000 44	
6	Wolf 359	M6V	2,700	0.000 021	
7	BD+36°2147	M2V	3,500	0.005 7	0.3
8	L 726–8A	M6V	3,000	0.000 059	0.044
9	L 726–8B	M6V		0.000 04	0.035
10	Sirius A	A1V	10,300	22.9	2.35
11	Sirius B		24,800	0.002 6	0.98
12	Ross 154	M4V	3,300	0.000 55	0.17
13	Ross 248	M5V	3,000	0.000 11	0.07
14	ε Eri	K2V	4,800	0.294	0.85
15	Ross 128	M4V	3,200	0.002 9	0.3
16	61 Cyg A	K4V	4,600	0.088	0.7
17	61 Cyg B	K5V		0.039	0.63
18	ε Indi	K3V		0.14	0.77
19	BD+43°44 A	M1V		0.006 1	
20	BD+43°44 B	M4V		0.000 39	
21	Luyten 789-6	M6V		0.000 14	
22	Procyon A	F5V	6,500	7.38	1.5
23	Procyon B		7,740	0.000 55	0.65
24	BD+59°1915 A	M3V		0.003 0	0.29
25	BD+59°1915 B	M4V		0.001 5	0.25

Next, here is a list of the 25 **visually brightest** stars in the sky, starting with the brightest (this list shares a few entries in common with the previous list):

	Star	Spectral Type	Temp.	Luminosity	Mass
A	Sun	G2V	5,800	1.0	1.0
B	Sirius A	A1V	10,300	22.9	2.35
C	Canopus	F0I–II	7,200	12,600	
D	α Cen A	G2V	5,800	1.60	1.10
E	Arcturus	K2III	4,500	150	1–1.5
F	Vega	A0V	10,800	50	2.6–3.1
G	Capella A	G1II		140	2.1
H	Rigel A	B8I	12,000	44,000	50
I	Procyon A	F5IV–V	7,000	7.6	1.7
J	Betelgeuse	M2I	3,300	8,700	
K	Achernar	B5V	18,800	1,260	
L	β Cen A	B1III		12,000	
M	Altair	A7IV–V	8,200	10.7	1.7
N	α Cru A	B1IV		3,500	
O	Aldebaran A	K5III	3,800	180	1
P	Spica	B1V	24,200	2,400	
Q	Antares A	M1I		13,000	
R	Pollux	K0III	4,500	30	
S	Formalhaut A	A3V		17	2.3
T	α Cen B	K0V	5,000	0.46	0.85
U	Deneb	A2I	9,700	48,000	25
V	β Cru	B0IV		5750	
W	Regulus A	B7V	13,000	145	3.5
X	Adhara A	B2II		8300	
Y	Castor A	A1		36.3	

On the next page is a graph showing each star's Temperature versus Luminosity. This is an **HR Diagram.** Data for the nearest stars have been plotted using a box symbol, while data for the brightest stars have been plotted using an "x". For the five stars that appear on both lists, they are plotted on the graph using both symbols, so you will see a box with a cross inside, which looks rather like a filled-in box.

Note that the tickmarks on the axes (shown on the top and sides) are not evenly spaced. Nonetheless, the small tick marks between each large tick mark represent equal sized numerical steps, e.g. on the vertical axis, between 1 and 10 are eight small tick marks, representing 2, 3, 4, 5, 6, 7, 8, and 9, and on the horizontal axis, between 3,000 and 10,000 there are six vertical dashed lines marking intervals of 1000, so representing 4000, 5000, 6000, 7000, 8000, and 9000. Moreover, between the dashed vertical lines for 3000 and 4000, there are nine small tick marks on the top axis (only) marking intervals of 100, so representing 3100, 3200, 3300... 3900.

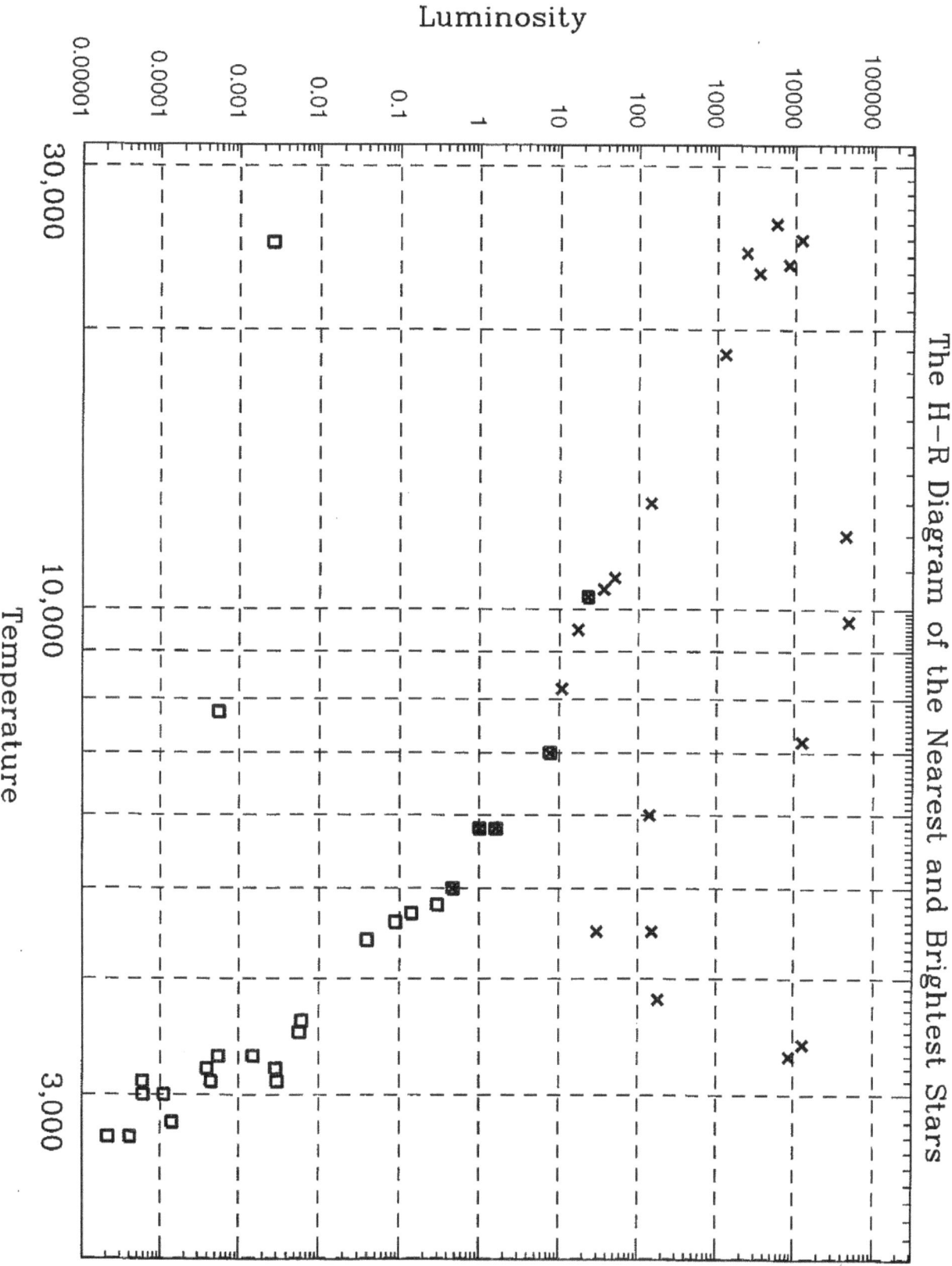

The H-R Diagram of the Nearest and Brightest Stars

Phase I: Exploration

Observe the general trend among a majority of the stars. This trend is called the "Main Sequence." *On the diagram above, circle the Main Sequence* -- that is, circle the stars which follow the trend that you observe; the stars are called "Main Sequence stars", and the trend is called the Main Sequence.

1. How many stars does the Main Sequence include for this particular diagram? _____ stars

2. Briefly describe the trend that defines the Main Sequence, making sure to include the properties of the stars that are represented and how they are related to one another.

3. If a star were measured to have a temperature of 3,500 K, predict, by examination of your HR Diagram, the luminosity that you think this star most likely has. Explain how you made your prediction, and any assumptions you made.

4. a) Find and **label by name** the following seven stars **on your graph**: The Sun, Barnard's Star, Sirius A, Sirius B, Regulus A, Deneb, and Betelgeuse.

4. b) Complete the following table indicating whether each star is on the Main Sequence.

Star's Name	The Sun	Barnard's Star	Sirius A	Sirius B	Regulus A	Deneb	Betelgeuse
On the Main Sequence?							

5. Stars in the top right corner of the graph are called Red Giants because their temperature is cool and thus their color is red, and they are swelled up to a huge "giant" size, making their luminosity very large. Which of those seven stars is closest to the top right corner of the graph?

6. Stars in the bottom left corner of the graph are called white dwarfs. They have high temperatures, glowing white or even blue, because they are the hot leftover cores of dead stars. Their sizes are very small though, "dwarfs" in fact, because they are undergoing no nuclear fusion to keep them swelled up. Their small size gives them a small luminosity. Which of those seven stars is closest to the bottom left corner of the graph?

7. For the seven Main Sequence stars you have already labeled on your graph, consult the tables at the beginning of the lab and write their mass next to their label *on the graph*.

8. Label the location <u>and</u> mass of the following additional four main sequence stars: L726-8A, Ross 154, 61 Cyg A, and Spica. Write them directly *on the graph*.

9. Rank the labeled Main Sequence stars (the preceding four plus the seven from the previous questions) <u>from highest to lowest</u> luminosity:

Highest luminosity   Lowest Luminosity

10. Describe any trend in the progression of masses that you just labeled along the Main Sequence.

Phase II – Does the Evidence Match the Conclusion?

The most luminous stars have either a very high temperature (hot things glow brightly), or a very large radius (the more glowing surface area there is, the more the total luminosity given off), or both. Consider the research question, "How does a star's radius change with its temperature and luminosity?" Use the following six hypothetical stars to explore this property of the HR Diagram.

Star name	Temperature (in K)	Luminosity (in $L_\odot$)	Radius
Alpha	3,000	.00001	the same as star Beta
Beta	30,000	.1	the same as star Alpha
Gamma	3,000	.001	10x bigger than star Alpha
Delta	30,000	10	10x bigger than star Alpha
Epsilon	3,000	.1	100x bigger than star Alpha
Zeta	30,000	1000	100x bigger than star Alpha

11. *Label the locations of these six* stars on the previous HR Diagram from Phase I.

12. Star Alpha and Star Beta have the same radius. *Connect the points representing Stars Alpha and Beta with a straight line using a straightedge.* This line represents a line of constant radius on the diagram. All stars on this line have the same radius!

13. Stars Gamma and Delta also have the same radius as one another. *Draw a straight line connecting these two stars.*

14. Stars Epsilon and Zeta have equal radii. *Draw a line connecting those two stars.*

15. You should now have three lines of constant radius on your HR diagram. Each line you drew represents a radius that is ten times larger than the previous line you drew. *Label the lines on your diagram accordingly, e.g. "Radius = 1 × Alpha," "Radius = 10 × Alpha," and "Radius = 100 × Alpha."*

16. The Sun should fall almost exactly along one of the lines you drew. How does the Sun's radius compare to that of Star Alpha?

17. Given the three lines drawn, what conclusions and generalizations can you make regarding the direction on the HR diagram in which the radii of stars increase?

18. *Continue drawing several more lines of constant radius (or size) on your HR diagram following the established pattern*, where each line represents a change of radius by a factor of 10 as before, until you have encompassed all the labeled stars on your diagram. (You will have to draw lines representing both larger <u>and</u> smaller radius than before.) *Label each line with the corresponding radius as before, relative to Star Alpha.*

19. Consider the research question **"How does the radius of a given star compare to the radius of the Sun?"** If a student claimed that the star Betelgeuse has a radius 500 times that of the Sun, would you agree or disagree? If a student claimed the Sun has a radius 500 times that of the star Sirius B, would you agree or disagree? *Explain your reasoning and provide specific evidence to support your claim.*

Phase III – What Conclusions Can You Draw From the Evidence?

Stars do not live forever. They are born and they die, sometimes passing through several different stages as they die, which causes their properties (namely Luminosity and Temperature) to change. If their properties change, their location on the HR Diagram changes. The one underlying property that determines all other properties a star will have -- and indeed dictates its whole life cycle -- is **mass**. The mass a star is born with determines what Luminosity and Temperature it will have (and therefore its place on the HR Diagram), what radius it will have, and how long it will live on the Main Sequence. Below is a table showing the Main Sequence lifetimes of stars of varying mass. Consider how star lifetime varies along the main sequence in terms of luminosity and temperature.

Star Mass (in solar masses)	Main Sequence Lifetime
17.5	12 million years
7.6	34 million years
5.9	100 million years
3.8	290 million years
2.9	540 million years
2.0	1.4 billion years
1.6	7 billion years
1.05	9 billion years
0.92	18 billion years
0.79	26 billion years

Stars often form in clusters, where a whole group of stars form at the same time. These stars form with a range of masses, spanning the whole Main Sequence. If you were to create an HR diagram of all the stars in a brand new cluster, all age zero ("0 Myr" means 0 million years), you should see the whole Main Sequence represented, as shown below. Keep in mind that a star's properties are constant during its Main Sequence lifetime, so a star stays in the same location on the HR Diagram during its whole Main Sequence lifetime, and only leaves the Main Sequence when it "dies."

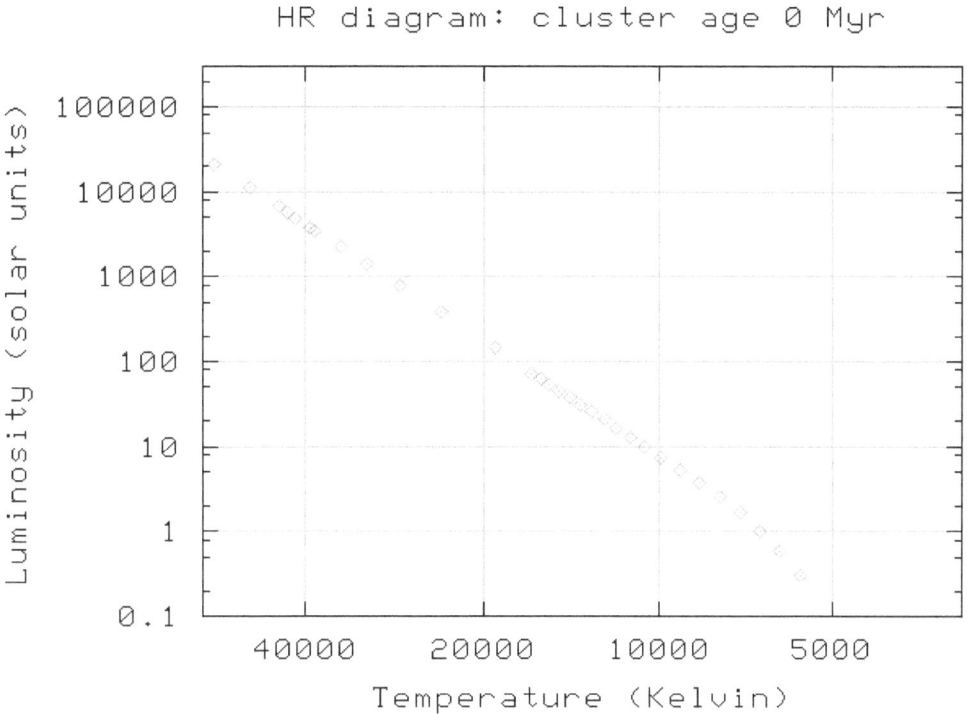

HR diagram: cluster age 0 Myr

Now, answer the research question, **"How would the main sequence of this HR diagram look in 5 billion years? 10 billion years? 25 billion years?"** *Explain your reasoning and provide evidence to support your conclusion.*

20. Evidence-based conclusion:

Phase IV – What Evidence Do You Need?

Most Main Sequence stars go through a sequence of phases like our Sun will: Main Sequence, then Red Giant, then White Dwarf. Stars are being born and dying all the time in an ongoing cycle, so at any given time, there are some stars in any of these stages. Your example HR Diagram shows a snapshot of what stage several dozen unrelated stars are in at one moment in time.

Imagine your team has been assigned the task of writing an article for your favorite science blog about the life cycle of Main Sequence stars. Describe precisely what evidence you would need to collect from an HR Diagram in order to answer the research question, **"Judging by what fraction of stars are in the Main Sequence, red giant, and white dwarf stages respectively, how long does a given star spend in each of those stages as it progresses through its life?"** You do not need to actually complete the steps in the procedure you are writing.

Hint: It might be instructive to think more about the analogy with people. If you sampled 50 random people from the population, what fraction of them would you expect to be school children? Working adults? Senior citizens? How does this relate to the relative length of time a given person spends in each of those phases?

21. *Create a detailed, step-by-step description of evidence that needs to be collected and a complete explanation of how this could be done—not just "look and see how many are in each stage," but exactly what would someone need to do, step-by-step, to accomplish this. You might include a table and sketches-the goal is to be precise and detailed enough that someone else could follow your procedure.*

Phase V – Formulate a Question, Pursue Evidence, and Justify Your Conclusion

Your task is design an answerable research question, propose a plan to pursue evidence, collect data using the HR diagram and included tables, and create an evidence-based conclusion about the characteristics of stars that you have not completed before.

Specific Research Question:

Step-by-Step Procedure, with Sketches if Needed, to Collect Evidence:

Data Table and/or Results:

Evidence-based conclusion Statement:

Phase VI – Summary

22. Create a 50-word summary, in your own words, that describes the properties of stars that can be represented on an HR Diagram, the trends in those properties across the graph, and the various groups of stars represented on an HR Diagram. You should cite specific evidence you have collected in your description, not describe what you have learned in class or elsewhere. Feel free to create and label sketches to illustrate your response.

23. Address how the distribution of the 25 brightest stars on the graph compares to that of the 25 nearest stars. Which group is more representative of a sample of stars in the galaxy as a whole, and why?

    Hint: Our galaxy contains a couple hundred billion stars with a whole range of properties spanning the whole HR Diagram. Some of these stars are nearby to us, but most are far away. Consider an example about people that may inform your evaluation of stars. Think about last time you were in a big crowd of many many people, say at a large sporting event. Let's say you made a list of the 25 fans sitting closest to your seat, and another list of the 25 fans that were the loudest you could hear. Which list of people, closest or loudest, would be more representative of most of the fans in the stadium?

www.ingramcontent.com/pod-product-compliance
Lightning Source LLC
Chambersburg PA
CBHW080639180526

45168CB00008B/3233